T0187846

MONOGRAPHS ON STATISTICS AND APPLIED PROBABILITY

General Editors

V. Isham, N. Keiding, T. Louis, N. Reid, R. Tibshirani, and H. Tong

Components
of Variance

D.R. Cox
Honorary Fellow
Nuffield College, Oxford, UK

P.J. Solomon
Associate Professor of Statistics
University of Adelaide, Australia

CRC Press
Taylor & Francis Group
Boca Raton London New York

CRC Press is an imprint of the
Taylor & Francis Group, an **informa** business

A CHAPMAN & HALL BOOK

CRC Press
Taylor & Francis Group
6000 Broken Sound Parkway NW, Suite 300
Boca Raton, FL 33487-2742

First issued in paperback 2019

© 2003 by Taylor & Francis Group, LLC
CRC Press is an imprint of Taylor & Francis Group, an Informa business

No claim to original U.S. Government works

ISBN-13: 978-1-58488-354-8 (hbk)
ISBN-13: 978-0-367-39597-1 (pbk)

Library of Congress Cataloging-in-Publication Data

Cox, D. R. (David Roxbe)
    Components variance / D.R. Cox and P.J. Solomon.
        p. cm. — (Monographs on statistics and applied probability ; 97)
Includes bibliographical references and index.
    ISBN 1-58488-354-5 (alk. paper)
1. Analysis of variance. I. Solomon, P. J. II. Title. III. Series.
QA279 .C69 2002
519.5′38—dc21

                                                                    2002073622

Library of Congress Card Number 2002073622

**Visit the Taylor & Francis Web site at
http://www.taylorandfrancis.com**

**and the CRC Press Web site at
http://www.crcpress.com**

Contents

Preface

The notion that haphazard variation may arise from a number of sources and that it may be valuable to identify these sources and measure their impact has a long history and many applications and implications. Indeed, it is only in very simple situations that it is likely to be satisfactory to represent haphazard variation by independent identically distributed random variables or by the essentially equivalent notion of random sampling from a hypothetical infinite population.

The statistical ideas, models and methods associated with structured haphazard variability arose in industrial applications especially in the textile industries via the work in the 1930s in the cotton industry by L.H.C. Tippett and in the wool industries by H.E. Daniels. In those industries producing very uniform output from very variable input is a key issue. It was in this setting that one of the authors (DRC) first encountered components of variance. The other author (PJS) met them in the context of the variation of blood pressure and other features in large clinical trials. There are, of course, many other applications, of which biometrical genetics, animal and plant breeding and psychometric testing are important examples.

Early work emphasized balanced data. More recently, driven largely by computational advances, the emphasis has shifted to unbalanced data and to the closely related theme of multi-level modelling in which the variation is attached to underlying parameters such as regression coefficients rather than directly to the observations.

In this book we have aimed to set out the essential principles of the subject concentrating on formulation of models which are the base for detailed analysis as well as on the statistical techniques themselves. A number of examples of realistic complexity are described at least in outline. The book is intended for a wide range of readers, some with no initial knowledge of the subject and others interested in very specific issues. To guide the reader towards key passages and particular topics we have preceded each chapter by a short preamble.

As in almost all applications of statistical methods computational aspects are both critical and subject to rapid change. We have aimed to give a guide to the current position without making the book too dependent on specific tools.

Although written primarily for statisticians, the book is intended to be accessible to users of statistical methods, although some familiarity with key theoretical concepts is needed for some parts, especially those describing new or recent developments.

The subject has a very large literature; in the bibliographic notes we have aimed to lead the reader to some of the key papers.

We are indebted to David Firth for his careful reading and comments on an earlier version of the manuscript, and to Michael Murray for his expertise in LaTeX.

We are grateful to The University of Adelaide and ARC (DRC) and to Nuffield College, Oxford and EPSRC (PJS) for support of travel in connection with our work.

D.R. Cox and P.J. Solomon

Oxford, UK and Adelaide, Australia

CHAPTER 1

Key models and concepts

Preamble

The chapter deals with simple situations involving more than one level of random variation. The emphasis is on careful specification of how the observations depend on underlying fixed parameters and random variables. There are two particularly important concepts. One is the distinction between nesting and cross-classification as aspects of data structure. This determines the nature of the sources of variation that can be examined from a particular set of data. The second is the choice between the representation of variation via fixed parameters. often as a basis for studying contrasts, and via random variables. usually summarised in the first instance by their variance. The second leads typically to consideration of a number of components of variance. For linear systems there are two parallel representations of the random part of the variation. one by sums of random variables and one by structured covariance matrices (Section 1.3). A number of examples are outlined. most notably aspects of the analysis of data from microarrays (Section 1.2).

1.1 Preliminaries

A recurring theme in statistical analysis concerns patterns of variation that are partly systematic and partly haphazard. The haphazard variation may arise from natural variability between similar individuals, be they patients. subjects, experimental animals. batches of material or whatever. or from measurement or sampling error. The systematic component may be variation between individuals to be explained via dependence on explanatory features or, in an experiment, via the experimental manipulations or treatments.

In the simplest situation of single observations on separate individuals the haphazard variation may be represented through independent random variables, one for each individual and. in particular, often by independent and identically distributed random variables. This leads us to various models that are widely applicable and in a sense the core of elementary statistical methods. Typical examples are to treat n observations of a response variable as having one of the following forms:

1

- $Y_j = \mu + \epsilon_j;$
- $Y_j = \beta_0 + \beta_1 x_j + \epsilon_j;$
- $Y_j = f(x_j, \theta) + \epsilon_j;$
- Y_j is binary, taking values 0 and 1 and with $P(Y_j = 1) = g(x_j, \theta).$

These represent, respectively, a random sample of observations, a linear regression on an explanatory variable x, a nonlinear regression on x of specified functional form involving unknown parameters θ and finally a series of binary observations with a distribution determined via the explanatory variable x and the function $g(x_j, \theta)$, and where $P(A)$ denotes the probability of the event A.

In the first three models the haphazard component is represented by independent and identically distributed random variables $\epsilon_1, \ldots, \epsilon_n$ of zero mean and variance $\tau = \sigma^2$, the random variables possibly also being assumed to be normally distributed. The other Greek letters represent unknown parameters or functions specifying the systematic structure assumed present.

All these models and their many immediate generalizations have a single random component corresponding to each observed individual. There are two rather different reasons why this may be inadequate. The first is that the haphazard variation may have more complex multi-component structure arising because that variation arises from several sources which can be identified. The second is that the parameters describing the systematic part of the variation may themselves change randomly, say between individuals or groups of individuals.

From a formal point of view these two possibilities may not need to be distinguished, although there are certainly differences in terms of the objectives of an analysis. The first approach has a long history under the general heading of *components of variance*. The second is frequently called *multi-level modelling* or sometimes *hierarchical modelling* or, in the older literature still *regression models of the second kind*. Many but certainly not all the former developments are tied to balanced data whereas the latter strand of work, much of it driven by the unbalanced observational data of the social sciences, is not restricted to balanced data and tends to make extensive use of powerful methods of fitting made feasible by modern computational developments.

We aim to cover both aspects of the topic although we begin with the former in which it is easier to see the conceptual issues involved largely shorn of technical detail.

The objectives of statistical analysis can be classified in various ways. For instance, one comparison is between analysis, interpretation and understanding as contrasted with prediction and the development of automatic or semi-automatic decision rules. In the present context, however, the main contrast is between interest lying primarily in the haphazard variability versus a focus on the systematic structure.

Illustration. A careful study of a measurement technique, for example in industrial chemistry, might take the following form. Several very different batches of raw material are individually thoroughly mixed and divided into subsamples as homogeneous as is feasible. Subsamples are sent to a number of labs at each of which there are a number of operators and sets of apparatus. The study protocol aims to achieve blindness subject to balance between batches, labs, operators and sets of apparatus. The object is to examine the relative contribution to variability of the various sources of haphazard variation, namely between subsamples, between sets of apparatus and between operators, allowing for possible interaction effects. The ultimate outcome may be suggestions for improved standardization and recommendations over the amount of replication desirable in routine application.

Illustration. In animal breeding studies, milk yields may be obtained from records in many farms, of different breeds of cattle, and of cows sired by various bulls, themselves classified in various ways. The objective is to assess the importance of the different sources of variability, in particular genetic components, to milk yield.

By contrast, the interest in other kinds of study may lie more in the nature of the systematic variation that is present.

Illustration. Suppose that during the progression of a disease a marker, for example log viral load in the case of AIDS, varies systematically for each individual patient. A first crude approximation might be a linear regression with time. Thus each patient has notionally parameters (β_0, β_1) in the notation above. In addition there will be random variation about the regression line. When a group of patients are examined, however, it would be very unlikely that they all have the same (β_0, β_1). One would first aim to explain that variation, in particular by a further higher level regression model in which (β_0, β_1) are themselves regarded as responses observed with an error of estimation. Even after this, however, some random variation would remain and therefore we consider models in which these parameters themselves have random structure. Interest focuses on the magnitudes of the random variation of individual responses about their regression line, in the variation in the intercepts and slopes, as well as, of course, on explanatory determinants of those regression parameters.

In the following treatment we begin by introducing some key ideas via simple cases where the random variation has a simple balanced structure with a small number of different components and lead on from this to more complex situations.

1.2 Some simple special models

Here are some important idealized situations.

Situation 1. A group of individuals each has a notional 'true' value of a feature, e.g., blood pressure. Let μ_j denote the value for individual j for $j = 1, \ldots, n_J$. For each individual one measurement is made by a conditionally unbiased method, i.e., for every μ_j the corresponding observation is

$$Y_j = \mu_j + \epsilon_j, \tag{1.1}$$

where

$$E(\epsilon_j) = 0, \quad \mathrm{var}(\epsilon_j) = \sigma_\epsilon^2 = \tau_\epsilon, \tag{1.2}$$

say. We call τ_ϵ the *component of variance* within individuals (or for sampling or measurement error or whatever).

The notion of 'true' value is, of course, not simple. It may be defined by a 'gold-standard' method of measurement different from that used in obtaining Y, in which case the assumption of no bias or systematic error is a serious one. In some situations, however, the 'true' value is the hypothetical mean of a large number of repetitions under the same conditions in which case the requirement that $E(\epsilon_j) = 0$ is a matter of definition.

Situation 2. Suppose that the individuals are regarded not as the basis for specific contrasts between individuals or groups of individuals but as a random sample from a hypothetical infinite population of mean μ. Then we write $\mu_j = \mu + \xi_j$, i.e.,

$$Y_j = \mu + \xi_j + \epsilon_j, \tag{1.3}$$

where by definition and the assumed unbiased character of the measuring process,

$$E(\xi_j) = 0, \quad \mathrm{var}(\xi_j) = \sigma_\xi^2 = \tau_\xi, \quad \mathrm{cov}(\epsilon_j, \xi_j) = 0. \tag{1.4}$$

Here τ_ξ is called the component of variance between individuals.

It follows that the variance has two components, one representing variability between individuals, the other representing variability within individuals, i.e.,

$$\mathrm{var}(Y_j) = \tau_\xi + \tau_\epsilon; \tag{1.5}$$

it is clearly not possible to estimate the separate components from data $Y_1, \ldots, Y_{n_J}$ without supplementary information. Such information might be repeated observations on an individual or an external estimate of τ_ϵ derived from other studies or from theory.

Situation 3. Suppose now that on each individual we make several measurements of each type in the context of Situation 2. This gives observations

$$Y_{js} = \mu + \xi_j + \epsilon_{js}, \tag{1.6}$$

where we shall assume that the repeat observations $s = 1, \ldots, n_S$ are *nested* within individuals, for example that observation 1 on individual j has no especial connection with observation 1 on individual k, $j \neq k$. Also we assume that all the random variables ϵ, ξ are mutually uncorrelated. this being a nontrivial assumption as applied to repeat observations on the same individual. Note that the correlation coefficient between repeat observations on the same randomly chosen individual is

$$\rho = \frac{\tau_\xi}{\tau_\xi + \tau_\epsilon}. \tag{1.7}$$

This is called the *intra-class correlation coefficient*.

With $n_S = 2$ as an example. the structure of the data may be represented as .

$$
\begin{array}{cccc}
\text{Individual} & 1 & Y_{11} & Y_{12} \\
 & 2 & Y_{21} & Y_{22} \\
 & 3 & Y_{31} & Y_{32} \\
 & \vdots & \vdots & \vdots \\
 & J & Y_{n_J 1} & Y_{n_J 2}
\end{array}
$$

The rows correspond to individuals $1. \ldots, n_J$ but the ordering of the observations within each row is arbitrary, and the second suffix does not carry subject–matter meaning across individuals. The model assumes the individual's condition is stable so that $\mu + \xi_j$ is constant, and that each observation has measurement error ϵ_{js}. where all the ϵ_{js}'s are uncorrelated.

This last assumption should be given reasonably critical consideration. Suppose, for example, that two samples of blood were taken from an HIV positive individual at the same clinic visit for the purpose of determining viral load. If the samples were combined and homogenized. split again into two subsamples and measured blind to their identity, then the measurement errors could reasonably be assumed to be uncorrelated. On the other hand in this configuration the magnitude of variation between distinct samples taken at the same time could not be assessed separately.

Generally, relevant aspects to be considered, many on points of design. include

- Is the assumption of uncorrelated errors reasonable? Is there some way of checking it?

- Was the investigation done 'blind'?

- How much of the variation is real variation between samples. and how much is local measurement error?

- How much better would we do in estimating the relative magnitudes of the between- and within-sample components of variance if there were. say, three replicates rather than two?

Table 1.1 *Percentage heroin measurements in nine illicit heroin preparations.*

Sample	% Heroin	
1	2.2	2.3
2	8.4	8.7
3	7.6	7.5
4	11.9	12.6
5	4.3	4.2
6	1.1	1.0
7	14.4	14.8
8	21.9	21.1
9	8.8	8.4

- What are the design implications for the estimation of the mean?

- What are the *relative* rather than absolute values of the variance components? We sometimes use relative values when, for instance, comparing different measuring techniques.

To illustrate this last point, if two ways of measuring some feature are studied on individuals drawn randomly from the same population but the measurements are not on directly comparable scales, the ratio of variance components between and within subjects would be a reasonable initial basis for comparison. On the other hand, if the two methods give measurements in the same units and two groups of subjects are used which may differ in their real variability, then the absolute rather than the relative values of components of variance should be used.

The following simple example illustrates several key features of the one-way component of variance model (1.6) described by Situation 3.

Example 1.1. Illicit heroin preparations. Nine samples of illicit heroin preparations were analysed in duplicate by gas chromatography (Skoog et al., 1995). The percentages of heroin obtained are given in Table 1.1.

The overall sample mean and standard deviation for the pooled data are 8.96% and 6.25%. However, assuming that the investigator analysed the duplicates from each preparation 'blind' and that the preparations can be regarded as a simple random sample from some population, the variability should be split into components between and within preparations, i.e., into τ_ξ and τ_ϵ, the latter representing measurement error. We will see how to estimate these variance components in Chapter 3. For now, note that the average difference in the replicate measurements is 0.042% which is only a small fraction of the observed standard deviation between the individual sample means of 6.44%. It is also apparent from Figure 1.1 that there is a

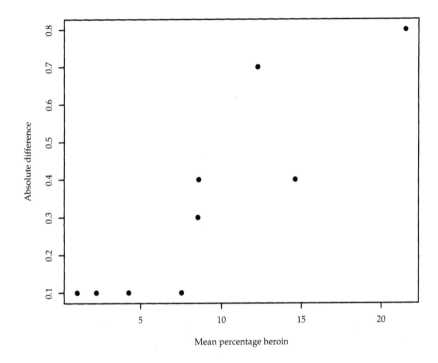

Figure 1.1 *Percentage heroin sample absolute differences versus means.*

variance–mean relationship in which the within-sample variance increases with the sample mean.

Situation 4. Suppose that individuals are cross-classified, in two directions. so that μ_{jk} is, for example, the 'true' mean for person j on visit k for $j = 1,\ldots,n_J; k = 1,\ldots,n_K$. Suppose also that there are observations nested within the individuals. For example, if Time 1 and Time 2 represent the beginning and end points of a study and there are $n_S = 3$ replicates nested within each patient $\times$ time combination, the structure of the data may be portrayed as in Table 1.2.

Suppose that, for this example, we are interested in this particular set of n_J patients, and the two particular times.

In general we write

$$
\begin{aligned}
Y_{jks} &= \mu_{jk} + \epsilon_{jks} \\
&= \bar{\mu}_{..} + (\bar{\mu}_{j.} - \bar{\mu}_{..}) + (\bar{\mu}_{.k} - \bar{\mu}_{..}) + (\mu_{jk} - \bar{\mu}_{j.} - \bar{\mu}_{.k} + \bar{\mu}_{..}) + \epsilon_{jks}.
\end{aligned}
\tag{1.8}
$$

Table 1.2 *Data layout for two-way cross-classified design given as an example of Situation 4.*

		Time 1	Time 2
Patient	1	Y_{111}	Y_{121}
		Y_{112}	Y_{122}
		Y_{113}	Y_{123}
	2	Y_{211}	Y_{221}
		Y_{212}	Y_{222}
		Y_{213}	Y_{223}
	$\vdots$	$\vdots$	$\vdots$
	n_J	$Y_{n_J 11}$	$Y_{n_J 21}$
		$Y_{n_J 12}$	$Y_{n_J 22}$
		$Y_{n_J 13}$	$Y_{n_J 23}$

in a conventional notation for averaging over suffices, so that, for example, $\bar{\mu}_{j.} = \Sigma_k \mu_{jk}/n_K$. We say that rows are crossed with columns and observations nested within row-column combinations. In some situations we might use the provisional working assumption that the row × column interaction term, the fourth on the right-hand side of (1.8), is zero.

Situation 5. Now suppose that one of the classifications, say rows, continues to represent individual levels of specific concern whereas the levels of the other, the columns, correspond to a random sample from an infinite population. For example, in a cross-over design, the rows may correspond to specific treatments of interest, whereas the columns correspond to patients. We then write

$$Y_{jus} = \mu + (\mu_j - \mu) + \xi_u + \eta_{ju} + \epsilon_{jus}, \tag{1.9}$$

where the ξ, η, ϵ are zero mean, uncorrelated random variables with variances $\tau_\xi, \tau_\eta, \tau_\epsilon$ representing components of variance between columns, for interaction and within cells. Note that in (1.8) $\bar{\mu}_{j.}$ is defined after averaging μ_{jk} over a finite set of levels of k, whereas in (1.9) μ_j is an average over a hypothetical infinite population of levels of (u, s).

Example 1.2. A crossover design for a heroin trial. A cross-over design was one of several proposed for a trial of prescription heroin in the Australian Capital Territory (Bammer and McDonald, 1994). The trial ultimately did not proceed following its rejection by the Federal Government in 1997, but in the cross-over design proposed, eligible injecting drug

Table 1.3 *Data layout for the* 3 × 3 *cross-over design for the heroin trial in Example 1.2.*

	Drug user			
	1	2	$\ldots$	n_J
Treatment				
IH	Y_{111}	Y_{232}	$\ldots$	$Y_{n_J 23}$
IH + OM	Y_{121}	Y_{212}		$Y_{n_J 33}$
OM	Y_{131}	Y_{222}		$Y_{n_J 13}$

users currently enrolled on a methadone program would have been randomized to one of three groups. The first group would have received injectable heroin (IH) for six months. followed by injectable heroin plus oral methadone (IH+OM) for six months. then oral methadone only (OM) for six months. The second group would have received treatments in the order IH + OM → OM → IH over the same 18-month period, and the third group OM → IH → IM + OM. It was also suggested that participants be allowed a free choice of any treatment for a further six months at the end of the trial, but we ignore that possibility here.

Numerous individual and community outcome measures were of interest. including licit and illicit drug use. health. social functioning and crime rates. Suppose Y measures duration of illicit drug use, then Table 1.3 sets out the hypothetical structure of the data assuming no replicates.

The first subscript, for example. indicates that individual 1 belongs to the first group which receives treatments in the order IH. IH+OM. OM. represented by the second subscript whereas the n_Jth individual belongs to the third group which receives treatments in the order OM. IH. IH+OM.

We can write the model as

$$Y_{jku} = \mu + \xi_j + \gamma_k + \delta_{ku} + \lambda_{k-1,u} + \epsilon_{jku} \qquad (1.10)$$

where ξ_j is a contribution specific to the jth subject. $j = 1.\ldots.n_J$: γ_k is the kth period parameter for $k = 1.2.3$: δ_{ku} is the direct effect of treatment in period k in the uth group, $u = 1.2.3$. and $\lambda_{k-1,u}$ is the carryover effect of treatment from period $k - 1$ in group u. Here, the specific treatment combinations are of interest, whereas the columns correspond to a sample of injecting drug users taken to be representative in some sense of the injecting drug population. The variance of Y_{jku} is $\tau_\xi + \tau_\epsilon$.

The investigators were keen to include a valid control group as a basis for comparison, here the 'oral methadone only' group, but there are obvious ethical and practical difficulties in implementing such a design. Switching individuals from one treatment regime to another after the relatively short period of six months could cause withdrawal problems as the effects of methadone are more long-lasting than those of heroin. Moreover, on the one hand, questions concerning long-term maintenance in treatment could not be addressed, and on the other hand, the time needed to measure outcomes reliably may be unacceptably long for participants. The treatments could not be given 'blind' and no allowance in this design is made for client choice. This last point is important as many drug users would be likely to consent to participate in such a trial in order to gain access to heroin. Compliance to treatment would therefore be a major design and analytical issue.

Situation 6. If rows are crossed with columns and observations nested within cells and both rows and columns are regarded as random samples from infinite populations, we write

$$Y_{jvs} = \mu + \xi_j + \eta_v + \zeta_{jv} + \epsilon_{jvs} \tag{1.11}$$

and so on.

Example 1.3. Spotted cDNA microarrays. Microarrays are a powerful tool for studying the expression levels of many thousands of genes simultaneously. They belong to the new biotechnologies designed to exploit DNA sequence data arising, for example, from the Human Genome Project, and are driving the new multidisciplinary fields of functional genomics and bioinformatics. The primary statistical challenge at present is the design, analysis and interpretation of the voluminous amounts of data representing the genetic blueprint of a living organism.

Spotted complementary DNA (cDNA) microarray experiments compare the relative quantities of messenger RNA sequences in different cell populations. In the first step of the technique, single-stranded DNA clones with known sequence content are robotically spotted out and fixed onto a glass slide or other solid substrate (i.e., the microarray). In the second step of the technique, purified mRNA from two cell populations under study are reverse-transcribed into cDNA and labelled with one of two fluorescent dyes, usually but not always, red and green. The two pools of differentially labelled cDNA are then combined and applied to the microarray during which process strands of cDNA from the pool hybridize to their complementary sequences on the array.

The fixed spots are usually called 'genes' although they can also represent expressed sequence tags (ESTs) of known or unknown function or DNA from another source. The red R and green G intensity signals from a spot represented by digital images measure the relative abundance of the mRNA in the two cDNA samples. Red is often used to label the 'treatment' sample, and green the 'control' sample, although increasingly so-called dye-swapped experiments in which the experiment is repeated with the dye-assignment reversed are used in practice to avoid confounding with differences between samples and to gain insight into an important class of systematic dye effects.

One of the primary aims of microarray experiments is to identify differentially expressed genes. The simplest experiments seek to identify changes in gene expression between different tissues types, drug treatments. or over time points in a biological process. A complete experiment will typically involve multiple slides and numerous sources of systematic and random variation which are not yet fully understood. The structure of the resultant data, the appropriate analyses and the quality and reliability of the results are determined by the experimental design as well as by the technical conduct of the experiment itself. In addition, the image processing for extracting information from the microarray images has a big impact on the downstream statistical analysis.

Two major issues of analytical importance for microarrays are adjusting the raw intensity measurements for *background,* and *normalisation.* The motivation for background correction is that a spot's measured fluorescence intensity for both R and G includes a contribution which is not due to the hybridization process of interest, but for example, may be due to the presence of other chemicals. We will assume here that the data are background adjusted and suitably transformed, usually to log base 2. being a natural scale of measurement for multiplicative changes in expression levels, commensurate with the 16-bit microarray images, and usually inducing effective additivity of the effects.

The purpose of normalisation is to identify and remove systematic and other sources of variation, especially biases due to different labelling efficiencies and scanning properties of the two dyes across genes. spatial effects on the slide, or other systematic experimental variation. A substantial proportion of the variation observed in microarray data is due to such systematic effects.

A relatively simple model which accommodates global normalisation effects and the gene-specific effects of interest in microarray experiments is of the form

$$Y_{jkuvs} = \mu + \xi_j + \kappa_k + \delta_u + \gamma_v + \lambda_{kv} + \zeta_{jvs} + \epsilon_{jkuvs}. \qquad (1.12)$$

where Y_{jkuvs} is the background-adjusted, log-transformed fluorescent intensity from array (i.e., slide) $j = 1, \ldots, n_J$, cDNA sample $k = 1. \ldots . n_K$. dye $u = 1, 2$, gene $v = 1, \ldots, n_V$ and gene replicate $s = 1. \ldots . n_S$. Recent experience has shown that the array and dye effects alone may not fully account for observed intensity and spatial dependence, and that it is advantageous to fit such models to the within-slide normalised intensities.

Using the Eisen and Brown (1999), Brown and Botstein (1999) protocols, between $n_V = 10,000$ and $n_V = 20,000$ cDNAs can be spotted onto a microscope slide. (The more technologically sophisticated Affymetrix arrays contain up to 40,000 distinct short oligonucleotides.) The number of samples or cell populations n_K typically range from 2 to 100 or more, depending on the nature of the investigation, and the actual number of arrays n_J may equal the number of samples n_K, or something larger if whole arrays are replicated. Here, n_S is the number of replicates of a particular gene within an array and will be equal to one or more depending on physical and other constraints, such as availability of mRNA.

The fixed gene effects of interest are the interactions λ_{vk} for gene v in cDNA sample k; the κ_k's are fixed main effects for cDNA samples which are of interest in their own right and occur when samples have higher or lower overall expression levels. The gene main effects, γ_v, occur when particular genes emit a higher or lower overall intensity compared to other genes, for example due to different labelling efficiency for different genes. The dye main effects, δ_u, measure differences in the two dye fluorescent labels (the green is usually brighter than the red, being a stronger dye), and μ represents the overall mean value. The array and spot effects ξ_j and ζ_{jvs} may be treated as random since overall differences in fluorescent signals from array to array can be viewed as having arisen from approximately normal distributions due to the accumulation of many small independent laboratory effects; ζ accounts for spot-to-spot variation within genes within arrays, here assuming that spots are replicated within arrays; ϵ is an independent random error term with mean zero and variance τ_ϵ.

Note that a more general model accounting for heterogeneity among genes would be to replace the independent error variance τ_ϵ by τ_{ϵ_γ}, and of course the model may be extended in other ways as appropriate.

Example 1.4. The inheritance of height. In statistical genetics, components of variance models are known as *polygenic models* and are widely used in animal and plant breeding. The term polygenic refers to the fact that the phenotype (i.e., the observed trait) is the result of the joint action of a very large number of genes, each with an individually small contribution to the phenotype, together with the environment. Such models are used in human genetics to help identify groups with a significant genetic component in their aetiology of disease, and in genetic counselling.

Height is widely cited as the classic polygenic trait and is typically modelled as the combined result of genetic and environmental components. Solomon et al. (1983) considered the form of the population distribution for height in 392 three-generation Finnish families. The primary aim of the early 1970s study from which these data were derived was to investigate genetic predisposition to high lipid levels and coronary disease, but height was found to constitute a major confounding factor.

A simple genetic model for an individual's height value adjusted for regional, generation and sex differences is

$$Y = \mu + \xi + \eta + \epsilon, \tag{1.13}$$

where Y is the phenotypic value, ξ is the additive genetic effect in the population representing the combination of main effects at all loci. η is the common sib-environment effect in the population which represents the resemblance between siblings reared together in a common environment. and ϵ is the individual environmental effect, i.e., residual variation. We include η here as an alternative to a dominant-type genetic effect as the two are indistinguishable on small pedigrees. Moreover, fixed regression effects z can be incorporated directly into the model via a term $\sum_k \beta_k z_k$ where the β's are regression parameters to be estimated.

Each of the random variables in (1.13) are assumed independently normally distributed with mean zero and components of variance τ_ξ. τ_η. τ_ϵ respectively, with the variance of Y equal to $\nu = \tau_\xi + \tau_\eta + \tau_\epsilon$. Covariance matrices of coefficients represent the degree of relationship between relative pairs of individuals j, k. An introductory exposition of the formulation of variance component models in terms of covariance matrices is given in the next section.

The basic polygenic inheritance model assumed for height in Solomon et al.'s study was

$$\xi_{\text{child}} = \frac{1}{2}(\xi_m + \xi_f) + \epsilon_p, \tag{1.14}$$

where m and f represent mother and father and ϵ_p denotes the individual genotypic variation, normally distributed with zero mean, that will be inherited by the offspring. Further. the correlation r between the observed heights of the parents, Y_m and Y_f. say. can be shown to induce covariance $r\tau_\xi H$, where $H = \tau_\xi/\nu$ is the *heritability* between ξ_m and ξ_f. giving the correlation between genetic effects in a couple. Heritability is an important quantity in population genetics and measures the proportion of the total phenotypic variance that is due to additive genetic variation.

Example 1.5. Woollen carding. In part of processing in the woollen industry, a wide web of material is divided into 100 ends of slubbing. arranged vertically in four large bobbins, each thus containing 25 ends. Uniformity in the mass per unit length of the product is crucial for the quality of the final output. To examine this, a typical study would weigh a fixed length of slubbing taken simultaneously from each bobbin. This then is repeated a number of times. The resulting data form a three-way cross classification arranged by times, bobbins and position across the card. The nature of the sources of variability that can be assessed from such data is set out in the following table, interactions between different sources. in the usual statistical meaning of interaction, being denoted by $\times$. The notion of an analysis of variance table will be developed in Chapter 2, but readers familiar with such tables may verify that if the number of repeat tests is three the degrees of freedom are as shown.

	Degrees of freedom
Bobbins	3
Across card	24
Times	2
Bobbins × Across card	72
Bobbins × Times	6
Across card × Times	48
Three-factor interaction	144

Each line in this table has an associated component of variance. The physical interpretation of each component centres on the technical process involved. One of the terms, that between times, is of little direct interest in the present context in that it reflects largely variation in the input material. Three of the terms represent systematic variation in principle capable of being eliminated or at least reduced by better adjustment of the machine. These are the variation between bobbins, the systematic variation across the card and the interaction between the two, sometimes called a tape effect, because it reflects systematic differences between the tapes that divide the material into the 100 ends. In addition to the bottom term which represents in some sense residual random variation, there are two sources of effectively random variation, namely interaction with times in the systematic effects across the card and between bobbins, representing what can only be treated as haphazard variation in the corresponding systematic effects. In reeling the fixed length for testing all ends from a given bobbin are dealt with together, so that any error in reeling contributes to all ends from the bobbin at that time and so is a contributor to the interaction Bobbins × Times.

1.3 A distributional specification

The development above is phrased in terms of underlying component random variables and this is the approach that in most respects is focused directly on the structure of the data and thus best for interpretation and as a base for generalization to new situations. Because the component random variables have zero mean, constant variances and zero correlations there is a virtually equivalent formulation in terms of covariance matrices. Also with the strong extra assumption that the component random variables are independently normally distributed, a full distributional specification is obtained.

To illustrate this we deal with the simple one-way model

$$Y_{js} = \mu + \xi_j + \epsilon_{js} \quad (j = 1, \ldots, n_J; s = 1, \ldots, n_S) \qquad (1.15)$$

where the ξ_j and ϵ_{js} have the properties previously indicated. For fixed j

we can regard the observations as a $n_S \times 1$ column vector with

$$\text{var}(Y_{js}) = \tau_\xi + \tau_\epsilon, \quad \text{cov}(Y_{js}, Y_{jt}) = \tau_\xi \ (s \neq t) \tag{1.16}$$

which can be written in alternative forms such as

$$\tau_\xi J_{n_S} + \tau_\epsilon I_{n_S} = (\tau_\xi + \tau_\epsilon) \mathcal{I}_{n_S}(\rho). \tag{1.17}$$

Here I_{n_S} and J_{n_S} are, respectively. the $n_S \times n_S$ identity matrix and the matrix all of whose elements are one. whereas $\mathcal{I}_{n_S}(\rho)$ is the corresponding intra-class correlation matrix all of whose off-diagonal elements are $\rho = \tau_\xi / (\tau_\xi + \tau_\epsilon)$.

Now under this specification

$$\text{var}(\Sigma_s Y_{js}) = n_S^2 \tau_\xi + \tau_\epsilon \{n_S + n_S(n_S - 1)\rho\}. \tag{1.18}$$

It follows that, excluding the degenerate case when ΣY_{js} is constant with probability one, we have that because the variance is positive

$$\rho > -1/(n_S - 1). \tag{1.19}$$

Thus the specification via the special form of the covariance matrix is slightly more general than the formulation with component random variables allowing, as it does, the formal possibility of a negative component of variance τ_ξ. Nevertheless for most purposes we regard the formulation in terms of random variables as primary.

The covariance matrix of the full $n_S n_J \times 1$ random vector formed by stacking the rows of $\{Y_{js}\}$ into a single column is a block diagonal matrix with elements (1.17), that is

$$(\tau_\xi J_{n_S} + \tau_\epsilon I_{n_S}) \otimes I_{n_J} = \tau_\xi U_\xi + \tau_\epsilon U_\epsilon. \tag{1.20}$$

where U_ξ, U_ϵ are associated matrices connected with indicator matrices defining the contribution of the component random variables to the observations. This formulation paves the way for a very general version with each separate component of variance identified with its own associated matrix.

In Chapter 4 we shall use the form of the covariance matrix to find the log likelihood function under the assumption that all component random variables are independently normally distributed. For this. however. we need not the covariance matrix but rather its inverse. In the simplest balanced cases this can be written down on noting. for example. that matrices of the form $aI_n + bJ_n$ have inverses of the same general form. forming what algebraists call a ring. In fact

$$(aI_n + bJ_n)^{-1} = a'I_n + b'J_n. \tag{1.21}$$

where

$$aa' = 1. \quad b' = -b/\{a(a + nb)\}. \tag{1.22}$$

1.4 Two key concepts

There are two quite different general ideas involved in the discussion so far. One is the distinction between nesting and cross-classification. This is concerned with the logical structure of the data and not with any specific probability model or distributional assumption. The other is concerned with whether we regard the levels of a 'factor' as intrinsically interesting or as a random sample from a population of levels, only the population features being of concern. Nested features will usually be treated this second way, and the distinction is often referred to as 'fixed' versus 'random' effects. Note especially, however, that where an analysis of variance table is involved, the form of the table is settled by the structure of the data and not by considerations of which contributions are to be regarded as random and which as fixed.

These two key concepts are subject-matter considerations. Generally, if the levels are treatments, they are fixed, although exceptions might include, for example, comparison of the effects of many different antibiotics or, in a plant breeding programme, of the properties of a large number of varieties of rice.

The formulation of a statistical model as a base for fruitful analysis thus depends crucially on the way the data were obtained as well as on more technical issues of independence, essential linearity of structure, etc., which in principle are usually capable of empirical test and which may need modification in any specific application.

1.5 Objectives

We shall consider statistical inference for components of variance or functions of them. The motivation may be

- intrinsic interest in the τ's as in comparison of different techniques and in genetics

- interest in the determinants of overall precision in estimating μ and other parameters

- design of further studies, in particular via *synthesis* of variance, i.e., recombination of the components under different assumptions or scenarios.

We consider first balanced situations in Chapters 2 and 3. Unbalanced situations involve more complicated estimation problems but are not otherwise essentially different, and we discuss these in Chapter 4. Chapter 5 concerns nonnormal data, beginning with the simplest Poisson and binomial models followed by extensions to survival data and more general situations. Finally, Chapter 6 is devoted to ways of extending and assessing models.

1.6 Bibliographic notes

While the idea of partitioning variability can be traced back at least to Airy's (1861) work on errors of observation in astronomy, systematic study especially related to R.A. Fisher's introduction of analysis of variance is more recent. Work in biometrical genetics can be traced via Bulmer (1980) and the systematic interest in the variability of industrial processes to Tippett (1931) and Daniels (1939). Eisenhart (1947) sets out very explicitly the distinction between fixed and random effect interpretations of an analysis of variance. The book of Searle et al (1992) gives a very systematic account largely of normal theory formulations and in particular provides a careful and lucid description of the matrix algebra underlying much of the discussion. C.R. Rao and Kleffe (1988) emphasize the point estimation of variance components using quadratic error loss. P.S.R.S. Rao (1997) gives a broad account of normal theory aspects.

For closely related ideas in psychometrics and educational statistics, see Birnbaum (1968).

Much of the initial work dealt primarily with balanced data. Henderson in a long series of papers starting with Henderson (1953), gave noniterative methods for unbalanced data based on equating suitable quadratic forms to their expectation; he was motivated by his studies of animal breeding. While these estimates are now largely replaced by likelihood-based methods, the more intuitively based methods still have some appeal. A general matrix formulation of the model for the unbalanced linear case leading to maximum likelihood estimation is due to Hartley and Rao (1967). The important generalization of maximum likelihood estimation to REML, reduced or restricted or residual maximum likelihood, is due to Patterson and Thompson (1971). Recent reviews with comprehensive bibliographies include Khuri and Sahai (1985) and an issue of *Statistical Methods in Medical Research* (Solomon, 1998) devoted to variance components.

One recent emphasis has been on situations requiring so-called multi-level modeling, in which underlying parameters themselves are subject to random variation. See, for example, Goldstein (1995) and Snijders and Bosker (1999). Verbeke and Molenberghs (2000) give an extremely thorough discussion of linear mixed models. Another has been on data requiring particular distributions such as the Poisson and binomial or on nonlinear normal theory models (Pinheiro and Bates, 2000).

For references on more detailed issues, see the separate chapters that follow.

Eisen and Brown (1999) is excellent introductory reading on microarrays, as is the series of articles in the Supplement to *Nature Genetics* (*The Chipping Forecast*, 1999). Two books edited by Schena (1999, 2000) provide general overviews of microarray technologies in addition to accessible accounts of different areas of application. There is a rapidly growing literature on statistical methods for the design and analysis of microarray experiments with an increasing emphasis on methods for identifying differentially expressed genes in multiple-slide experiments. Early work on identifying correlated cluster profiles focused on exploratory cluster analysis tools (Alizadeh et al., 2000) and there has been new work based on such approaches designed specifically for microarray experiments, for example, gene shaving (Hastie et al., 2000). However, relatively well-established statistical modelling, estimation and testing are increasingly being recognized as providing a sound basis for statistical inference and are often preferable. See Dudoit et al. (2002) for a general overview and cogent arguments for assessing gene significance from univariate testing procedures adjusted in appropriate ways for multiple testing. Kerr and Churchill (2001), Kerr et al. (2000) advocate orthodox experimental design and analysis of variance models similar to the prototype model in Example 1.3, but rather assume fixed effect linear models for the log intensities. Wolfinger et al. (2001) fit sequential 'normalisation' and 'gene' models, where the gene model is fitted to the residuals estimated from the global normalisation model. Jin et al. (2001) present a major application of mixed models to the analysis of microarray data. An empirical Bayes approach is described in detail in Chapter 3.

T.P. Speed's bioinformatics website `http://www.stat.berkeley.edu/users/terry/zarray` has links to major research being undertaken in microarrays and bioinformatics around the world. Another very useful website compiled by G. Smyth is at `http://www.statsci.org/micrarra/index.html`. The Bioconductor Project `http://bioconductor.org` is an international collaborative network and archive for the design and analysis of microarray experiments and bioinformatics more generally. It is based in the Biostatistics Unit of the Dana Farber Cancer Institute, Harvard Medical School/Harvard School of Public Health. The Bioconductor website contains useful links and software based on the free statistical and graphics package R.

Hopper (1993) provides a useful overview of statistical genetics up to the early 1990s. More recent developments are summarized in Balding et al. (2001) and in the encyclopedia of statistical genetics edited by Elston et al. (2002).

The example of variability in a complex industrial process is described in more detail by Daniels (1938, 1939). Many studies of this kind were done in the textile industries around that time but the ideas involved have not lost their relevance.

1.7 Further results and exercises

1. With some data from a problem like Situation 3, i.e., between and within variation with replication, it is found that the variation between

individuals is *much* less than that within individuals. What are the possible interpretations?

2. Show that in Situation 3 the correlation coefficient between repeat observations on the same individual is the intra-class correlation coefficient

$$\rho = \frac{\tau_\xi}{(\tau_\xi + \tau_\epsilon)}.$$

How would it be estimated by a formula analogous to that for an ordinary correlation coefficient?

3. Repeat observations are made on two (or more) groups of independent individuals using the same method. Discuss whether intra-class correlation or variance components are the best base for interpretation.

4. In the simple one-way arrangement the correlation coefficient between a single observation Y_{js} on an individual and the mean $\bar{Y}_{j.}$ is in psychometrics called Cronbach's alpha. Show that it is equal to

$$\sqrt{\{(\tau_\xi + \tau_\epsilon/n_S)(\tau_\xi + \tau_\epsilon)^{-1}\}}.$$

It is sometimes stated that values of α greater than or equal to 0.6 are satisfactory. Show that provided n_S is not too small this rule of thumb is equivalent to requiring the standard deviation between individuals to be more than three-quarters times the standard deviation within individuals.

5. In the system $Y_{js} = \mu + \xi_j + \epsilon_{js}$. show that

(a) one observation on an individual chosen at random from the population of individuals has variance $\tau_\xi + \tau_\epsilon$:

(b) two repeat observations on the same individual have a mean square difference of $2\tau_\epsilon$;

(c) if two individuals are chosen at random and respectively r_1 and r_2 independent observations made on them. then the variance of the difference of the means is $2\tau_\xi + \tau_\epsilon(1/r_1 + 1/r_2)$;

(d) prove that the requirement that $E(\epsilon_{js}) = 0$ conditionally on $\mu + \xi_j$ implies that $\mathrm{cov}(\xi_j, \epsilon_{js}) = 0$. Discuss some special cases in which this condition is satisfied without ξ_j and ϵ_{js} being independent.

6. It is known that a correlation coefficient R from a random sample of size n from a bivariate normal distribution of correlation coefficient ρ is such that $Z = \frac{1}{2}\log\{(1 + R)/(1 - R)\}$ is nearly normally distributed with mean $\zeta = \frac{1}{2}\log\{(1 + \rho)/(1 - \rho)\}$ and variance $1/(n - 3)$. Suppose that $R_1, \ldots, R_{n_J}$ are independent such correlations with corresponding parameters $\zeta_1, \ldots, \zeta_{n_J}$ which are themselves distributed with mean μ_ζ and variance τ_ζ. Show that the one-way representations can be applied with the lower component of variance known. What additional analyses would be sensible to interpret a non-zero value of τ_ζ?

CHAPTER 2

One-way balanced case

Preamble

For balanced systems of the type sketched in Chapter 1 there are parallel decompositions of the data vector, of sums of squares of the components and of the dimensions (degrees of freedom) of the spaces over which the components vary (Section 2.1). These are conventionally summarised in an analysis of variance table. There is no necessary probabilistic element in this; the form of the table indicates within a given framework what can be estimated from specified data. This leads under the kind of model discussed in Chapter 1 to a variety of statistical procedures. For balanced data and under normal-theory assumptions standard procedures based on the Student t and F distributions are briefly described (Section 2.2.1); for more general problems the idea of effective degrees of freedom is discussed (Section 2.2.2). Synthesis of variance and the use of variance components in planning sampling schemes are discussed (Section 2.3). The special problems of components of variance for a finite population are described (Section 2.4). Key issues of formulation are summarised (Section 2.5). Some introductory comments are made about the use of the method of maximum likelihood (Section 2.6).

2.1 Analysis of variance

In one sense the whole objective of the study of components of variance is analysis, i.e., the splitting up, of variance into parts. Usually, though, the term *analysis of variance* is used more narrowly to mean procedures based on the following decompositions which we illustrate first via Situation 3. i.e., model (1.6) for between and within variation with replication.

We decompose

1. The observation vector into orthogonal components, namely

$$Y_{js} = \bar{Y}_{..} + (\bar{Y}_{j.} - \bar{Y}_{..}) + (Y_{js} - \bar{Y}_{j.});\tag{2.1}$$

2. The corresponding sums of squares, namely

$$\Sigma Y_{js}^2 = \Sigma \bar{Y}_{..}^2 + \Sigma(\bar{Y}_{j.} - \bar{Y}_{..})^2 + \Sigma(Y_{js} - \bar{Y}_{j.})^2.\tag{2.2}$$

where all sums are over *all* suffices; note, for example, $\Sigma_{j,s}(\bar{Y}_{j.} - \bar{Y}_{..})^2$ is often written $n_S \Sigma_j(\bar{Y}_{j.} - \bar{Y}_{..})^2$;

21

3. The corresponding dimensions (degrees of freedom) in which the component vectors live, namely

$$n_J n_S = 1 + (n_J - 1) + n_J(n_S - 1). \tag{2.3}$$

It is helpful to write out the observation vector decomposed into three orthogonal vectors according to the algebraic identity in (2.1). Thus for $n_J = 3$ individuals and $n_S = 2$ observations nested within each individual, all the observations may be written as a vector (the ordering does not matter, but it may as well be sensible) decomposed as follows

$$
\begin{pmatrix} Y_{11} \\ Y_{12} \\ Y_{21} \\ Y_{22} \\ Y_{31} \\ Y_{32} \end{pmatrix}
=
\begin{pmatrix} \bar{Y}_{..} \\ \bar{Y}_{..} \\ \bar{Y}_{..} \\ \bar{Y}_{..} \\ \bar{Y}_{..} \\ \bar{Y}_{..} \end{pmatrix}
+
\begin{pmatrix} \bar{Y}_{1.} - \bar{Y}_{..} \\ \bar{Y}_{1.} - \bar{Y}_{..} \\ \bar{Y}_{2.} - \bar{Y}_{..} \\ \bar{Y}_{2.} - \bar{Y}_{..} \\ \bar{Y}_{3.} - \bar{Y}_{..} \\ \bar{Y}_{3.} - \bar{Y}_{..} \end{pmatrix}
+
\begin{pmatrix} Y_{11} - \bar{Y}_{1.} \\ Y_{12} - \bar{Y}_{1.} \\ Y_{21} - \bar{Y}_{2.} \\ Y_{22} - \bar{Y}_{2.} \\ Y_{31} - \bar{Y}_{3.} \\ Y_{32} - \bar{Y}_{3.} \end{pmatrix}.
$$

Note that the cross-product terms vanish verifying the orthogonality of the three component vectors.

In general, the number of possible dimensions in which the observations can vary is $n_J n_S$, which is split geometrically as follows: there is one degree of freedom for the mean (because if any one value in the mean vector is known, they are all known); there are n_J values in the vector of deviations about the overall sample mean, but by knowing $n_J - 1$ values, the n_Jth can be determined; and for each j, knowing $n_S - 1$ values determines the last value, leading to $n_J(n_S - 1)$ degrees of freedom for the within-individual component.

We call the sums of squares in (2.2) by the obvious names: Mean, Between individuals, and Within individuals, and define mean squares MS_ξ, MS_ϵ, as sums of squares divided by the corresponding degrees of freedom, d_ξ, d_ϵ. Note that this and similar decompositions are geometrical properties of the structure of the data and probability is not specifically involved.

These results can be summarised in the analysis of variance table given by Table 2.1.

From now on, summary analyses such as Table 2.1 are given in a slightly modified form in which the term for the mean is omitted from the table and the term then called Total is the sum of squares of deviations from the overall mean, namely $\Sigma Y_{js}^2 - \Sigma \bar{Y}_{..}^2$.

Now bring in the probability model $Y_{js} = \mu + \xi_j + \epsilon_{js}$. To illustrate the types of calculations employed here, consider the model for a particular individual j and average over the suffix s. The structure for the individual sample mean is

$$\bar{Y}_{j.} = \mu + \xi_j + \bar{\epsilon}_{j.}. \tag{2.4}$$

Table 2.1 *Analysis of variance table for the one-way layout of Situation 3.*

	SS	df	MS
Mean	$\Sigma \bar{Y}_{..}^2$	1	
Between individuals	$\Sigma(\bar{Y}_{j.} - \bar{Y}_{..})^2$	$n_J - 1$	MS_ξ
Within individuals	$\Sigma(Y_{js} - \bar{Y}_{j.})^2$	$n_J(n_S - 1)$	MS_ϵ
Total	ΣY_{js}^2	$n_J n_S$	

where the variance of $\bar{\epsilon}_{j.}$ is τ_ϵ/n_S. Then $E(\mathrm{MS}_\xi) = n_S\tau_\xi + \tau_\epsilon$. Since $E(\mathrm{MS}_\epsilon) = \tau_\epsilon$, this tells us how to estimate τ_ξ.

In summary, the key properties which follow directly from the properties of estimated variances from a simple random sample are

1. $E(\mathrm{MS}_\epsilon) = \tau_\epsilon$;

2. Each mean $\bar{Y}_{j.}$ has a random component $\xi_j + \bar{\epsilon}_{j.}$ of variance $\tau_\xi + \tau_\epsilon/n_S$:

3. $E(\mathrm{MS}_\xi) = n_S\tau_\xi + \tau_\epsilon$;

4. Hence τ_ξ is estimated via $(\mathrm{MS}_\xi - \mathrm{MS}_\epsilon)/n_S$:

5. $E(\bar{Y}_{..}) = \mu$ and $\mathrm{var}(\bar{Y}_{..}) = (n_S\tau_\xi + \tau_\epsilon)/(n_J n_S)$;

6. Therefore

$$(\bar{Y}_{..} - \mu)/\{\mathrm{MS}_\xi/(n_J n_S)\}^{1/2}$$

is a pivot for estimation of μ.

The resulting estimates are sometimes called the *least-squares-based* estimators and are unbiased estimates of the variance components. In Chapter 4, we compare these with the (biased) maximum likelihood estimates.

Assuming the pivot is approximately $N(0.1)$. we can obtain confidence limits for μ. For example. an approximate 95% confidence interval is

$$\bar{Y}_{..} \pm 1.96\sqrt{\left(\frac{\mathrm{MS}_\xi}{n_J n_S}\right)}. \tag{2.5}$$

These and very similar formulae are the immediate basis of the analysis of balanced situations and more complicated extensions apply to unbalanced cases. In a more refined version the $N(0.1)$ pivotal distribution is replaced by a Student t distribution with $n_J - 1$ degrees of freedom. For explicit justification of the Student t distribution we need the much stronger assumption that the underlying random variables are independently normally distributed. The use of a distribution longer tailed than the normal distribution is appropriate somewhat more generally. however.

Example 2.1. Illicit heroin preparations continued. Recall from Chapter 1 that two replicate measurements were made on each of nine preparations.

The simple one-way component of variance model allows the standard deviation of the pooled sample, previously obtained as 6.25%, to be split into components of standard deviation between and within samples.

The between-sample standard deviation is the dominant source of variability and is estimated to be 6.44% with the standard deviation of measurement error being 0.30%. Thus the total standard deviation of the response is estimated to be 6.45%. Comparing this to the estimated pooled standard deviation of 6.25% found in Chapter 1, we can see that by not accounting for the error structure in the appropriate way, the total variation in the measurements is under-estimated.

Example 2.2. The IPPPSH. The International Prospective Primary Prevention Study in Hypertension was a large-scale multi-centre clinical trial designed to compare the efficacy of combination therapies in preventing heart disease and stroke (IPPPSH Collaborative Group 1984, 1985). Between 1977 and 1980, 6,357 men and women from six countries aged 40 to 65 years and with uncomplicated essential hypertension were randomized to receive treatment regimes with or without the beta-blocker oxprenolol. Patients were followed-up at quarterly intervals for periods ranging from three to five years.

Solomon (1985) and Solomon and Cox (1992) have previously analysed quarterly blood pressure measurements on 25 men on oxprenolol from the four-year cohort, for which diastolic and systolic blood pressures were measured (in mmHg) in duplicate at each quarterly clinic visit. Diastolic blood pressure was subject to a target control level of < 95 mmHg, but systolic blood pressure was not specifically targetted in this way.

Figure 2.1 depicts the diastolic and systolic blood pressure trajectories for each patient. There is an overall tendency for diastolic pressure to decline or to remain relatively stable, which is expected in light of the target control level. There is also an appreciable amount of variation from visit to visit within individuals. The patterns of variability within and between patient systolic blood pressure measurements typically, although not uniformly, mirror those for diastolic pressure. Again it is expected that systolic pressure will, to an extent, reflect changes in diastolic pressure at an individual level. Nonetheless, systolic measurements within and between individuals are seen to be considerably more variable.

A natural hierarchical model for considering diastolic or systolic blood pressure measurements separately, ignoring for the moment possible trends or serial correlation, is to split the variation into components between patients, between visits within patients (representing long-term variability) and between replicate measurements within patient visits, the latter representing measurement error or short-term variability. We write

$$Y_{jks} = \mu + \xi_j + \eta_{jk} + \epsilon_{jks} \tag{2.6}$$

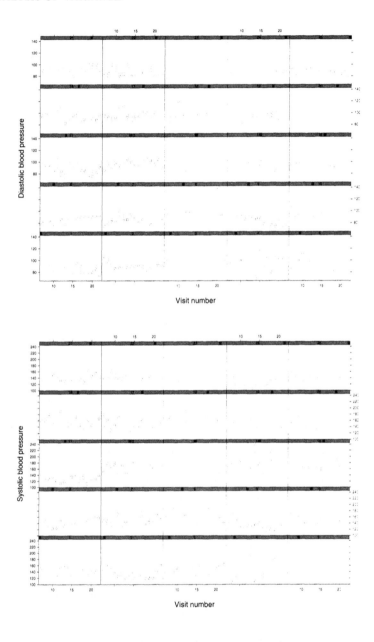

Figure 2.1 *Diastolic (above) and systolic (below) blood pressures in mmHg taken twice at quarterly intervals over four years for 25 men numbered 1 to 25. Each patient retains the same position in the two grids of plots. The horizontal axis shows the visit number (7 to 16) where 7 is the first visit following the stabilization of the patient on anti-hypertensive treatment. On the vertical scale, the observed diastolic blood pressures range from 68 to 140 mmHg, and the observed systolic pressures range from 105 to 240 mmHg.*

Table 2.2 *General form of the analysis of variance table for the hierarchical model with three nested effects describing the IPPPSH blood pressure data analysed in Example 2.2. BP: between patients; BV: between visits within patients; W: between replicates within patient visits.*

	SS	DF	$E(\text{MS})$
BP	$\Sigma(\bar{Y}_{j..} - \bar{Y}_{...})^2$	$n_J - 1$	$\tau_\epsilon + n_S(\tau_\eta + n_K\tau_\xi)$
BV	$\Sigma(\bar{Y}_{jk.} - \bar{Y}_{j..})^2$	$n_J(n_K - 1)$	$\tau_\epsilon + n_S\tau_\eta$
W	$\Sigma(\bar{Y}_{jks} - \bar{Y}_{jk.})^2$	$(n_S - 1)n_Jn_K$	τ_ϵ

Table 2.3 *Analysis of variance table for the IPPPSH diastolic blood pressure data. BP: between patients; BV: between visits within patients; W: between replicates within patient visits.*

	SS	DF	MS	$E(\text{MS})$
BP	23058.66	24	960.78	$\tau_\epsilon + 2\tau_\eta + 32\tau_\xi$
BV	39081.94	375	104.22	$\tau_\epsilon + 2\tau_\eta$
W	3092.00	400	7.73	τ_ϵ

for $j = 1, \ldots, 25; k = 1, \ldots, 16$ and $s = 1, 2$; μ is the overall mean and a constant; and the random variables ξ, η and ϵ are mutually independent with variances τ_ξ, τ_η and τ_ϵ. This extends Situation 3 for between and within variation with replication by adding a random term.

For the general balanced case with n_J patients, n_K visits within patients, and n_S replicates within patient visits, the appropriate analysis of variance table, generalizing that for the one-way model, is given by Table 2.2.

Note that $E(\bar{Y}_{...}) = \mu$ and, as before,

$$\text{var}(\bar{Y}_{...}) = \tau_\xi/n_J + \tau_\eta/(n_Jn_K) + \tau_\epsilon/(n_Jn_Kn_S). \tag{2.7}$$

For the IPPPSH diastolic blood pressure data, we obtain Table 2.3.

From Table 2.3 we estimate the variance components between patients, between visits within patients and for replicates to be 26.77, 48.25 and 7.73 mmHg^2, respectively. The between visit component of variance makes the biggest contribution to the observed variability. These results are in close agreement with those obtained by maximum likelihood based on a bivariate nested model. In general, minor discrepancies are likely to arise from rounding errors from different computer programs. The values in the table were obtained using `raov` in S-PLUS and to two decimal places the estimated variance components are virtually identical to those obtained from `lme`, also in S-PLUS.

Solomon (1985) established via formal maximum likelihood that blood pressure data are close to log normally distributed. We discuss this and transformations more generally in Chapter 3.

2.2 Some more assumptions

For more formal properties the simplest further assumptions are that all random variables are independently normally distributed. Of course this is a working simplifying assumption to be made neither uncritically nor universally. Later we shall discuss ways of investigating the adequacy of the assumption but often the best strategy is, within reason, to assume normality and then to consider how sensitive the conclusions are likely to be to that assumption. In particular instances, specific non-normal distributions may be proposed for one or both components.

2.2.1 Some exact theory: key results

Under normality the following properties hold for Situation 3, between and within variation with replication. Extensions to general balanced cases are straightforward.

1. The two sums of squares and the sample mean are minimal sufficient statistics implying various strong optimum properties. This provides a very strong consequence of assuming normality. For as long as the model is reasonably adequate, all that are needed are the sums of squares and the mean.

2. The two sums of squares are independently distributed proportionally to chi-squared variables with the appropriate degrees of freedom.

3. The following pivots thus allow formally exact inference for the relevant parameters:

$$d_\epsilon \mathrm{MS}_\epsilon / \tau_\epsilon; \tag{2.8}$$

$$d_\xi \mathrm{MS}_\xi / (\tau_\epsilon + n_S \tau_\xi); \tag{2.9}$$

$$(\mathrm{MS}_\xi / \mathrm{MS}_\epsilon)(1 + n_S \tau_\xi / \tau_\epsilon)^{-1}; \tag{2.10}$$

$$(\bar{Y}_{..} - \mu) / \{\mathrm{MS}_\xi / (n_J n_S)\}^{1/2}. \tag{2.11}$$

The pivotal distributions are chi-squared for the first two, F with degrees of freedom (d_ξ, d_ϵ) and Student t with degrees of freedom d_ξ.

4. In particular we could, if we wished, test the typically uninteresting hypothesis $\tau_\xi = 0$.

5. Only particular parametric functions have 'exact' confidence intervals.

2.2.2 Some slightly less simple aspects

The above results lead to simple answers to a limited range of questions, for example to confidence limits for τ_ξ/τ_ϵ but not for τ_ξ itself. Further if we have several independent similar sets of data we may want to combine estimates across sets of data, subject to checks of homogeneity.

The safest general procedure for doing this is the use of profile likelihood or one of its modifications. For many, but not all, purposes the following device is much simpler and essentially equivalent.

We note the following:

1. If Z is distributed proportionally to a chi-squared variable with d degrees of freedom, then

$$d = 2\{E(Z)\}^2/\text{var}(Z). \qquad (2.12)$$

2. If we have a reasonable estimate T for the parameter θ and T is either always positive or at least has very high probability of being positive, it may be reasonable to approximate its distribution by an appropriate multiple of chi-squared with d degrees of freedom, with d given in (2.12). Nearly always d has to be replaced by an estimate $\tilde{d}$. Thus if T is exactly or approximately unbiased for θ it can be treated as a standard estimate of variance with *effective degrees of freedom* $\tilde{d}$. In general $\tilde{d}$ will not be an integer and interpolation will be needed to obtain percentage points.

3. The method is likely to fail badly if negative values of T have appreciable probability.

For example $(\text{MS}_\xi - \text{MS}_\epsilon)/n_S$ is an unbiased estimate of τ_ξ. It will often have essentially positive support. Because of the chi-squared forms for the two mean squares it follows that

$$\tilde{d} = (\text{MS}_\xi - \text{MS}_\epsilon)^2/(\text{MS}_\xi^2 d_\xi^{-1} + \text{MS}_\epsilon^2 d_\epsilon^{-1}). \qquad (2.13)$$

This is an important general technique for problems that do not fall into direct inference about τ_ξ and τ_ξ/τ_ϵ.

Tests of homogeneity and the pooling of suitable information from related studies are in principle best done by likelihood-based procedures. In many applications, however, the following is a much simpler more transparent and essentially equivalent approach. If T is an approximately unbiased estimate of an essentially positive parameter θ with degrees of freedom, 'exact' or effective, d, then $\log T$ is approximately normally distributed around mean $\log \theta$ with variance $2/d$. See also Section 6.6. This throws further issues of analysis into a normal theory least squares framework.

For example, suppose we want to know whether τ_ξ is the same across several studies. The mean squares MS_ξ will be different, and the τ_ϵ, n_J and n_S may also differ. The procedure is then to estimate τ_ξ for each study, take the log $\hat{\tau}_\xi$ and treat them as a sample with variance $2/\tilde{d}$. It is also possible of course to relate τ_ξ to explanatory variables.

It can be shown that if a standard estimate of variance with d degrees of freedom is found when the underlying observations are independently distributed with fourth cumulant ratio γ_2 then approximately the effective degrees of freedom are $d/(1+\gamma_2/2)$. This result can be used in at least two ways. First it can be used as a basis for a sensitivity analysis to assess what level of nonnormality would have to be present to make a major change in the conclusions from a particular analysis. Secondly an estimate of the fourth cumulant ratio can be used to construct a modified set of confidence intervals or test adjusting for nonnormality. A difficulty with the latter procedure is that estimates of fourth cumulants are notoriously unstable.

2.2.3 Negative estimates of components of variance

The variances τ are nonnegative. On the other hand the standard estimates of upper variance components are based on differences of mean squares and hence may sometimes be negative. The simplest and most generally satisfactory way to deal with this is to replace notionally negative values by zero. For instance in the simplest one-way model (Situation 3 for between and within variation with replication) we take as estimate of τ_ξ

$$\max\{(\mathrm{MS}_\xi - \mathrm{MS}_\epsilon)/n_S, 0\}. \tag{2.14}$$

There are two qualifications. First, if MS_ξ is substantially less than MS_ϵ, in particular if MS_ξ is significantly less than MS_ϵ as judged by an F test, the data are inconsistent with the postulated model. This is typically warning that a systematic effect has been omitted from the original model, for example that data that should have been treated as a two-way arrangement have been collapsed into a one-way analysis. An alternative somewhat similar explanation is in terms of neglected correlations between the relevant random variables.

Secondly if a composite estimate of, say τ_ξ is to be produced by averaging from a number of separate sets of data, then negative values should be preserved in the original estimates to avoid systematic error in the pooled estimate.

A rather more formal justification of (2.14) can in effect be obtained if we regard the primary mode of statistical inference the provision of confidence limits for unknown parameters of interest. Thus if, to show the argument in its clearest form, we take τ_ξ/τ_ϵ as the parameter of interest then

$$P\{F_\alpha < (\mathrm{MS}_\xi/\mathrm{MS}_\epsilon)(1 + n_S\tau_\xi/\tau_\epsilon)^{-1} < F_{1-\alpha}\} = 1 - 2\alpha. \tag{2.15}$$

where $F_\alpha, F_{1-\alpha}$ are quantiles of the standard F distribution. Therefore formally a $1 - 2\alpha$ confidence interval is

$$F_\alpha^{-1}\mathrm{MS}_\xi/\mathrm{MS}_\epsilon - 1 > \tau_\xi/\tau_\epsilon > F_{1-\alpha}^{-1}\mathrm{MS}_\xi/\mathrm{MS}_\epsilon - 1. \tag{2.16}$$

Now suppose that any negative limits in the above statement are replaced by zero. The coverage property will be unchanged for any nonzero value of the parameter. If $\tau_\xi = 0$ the coverage probability will be increased. This provides some formal justification for the common-sense procedure. For parameter combinations for which formally exact inference is not available, an approximate version of the above applies.

2.3 Synthesis of variance

A common application of components of variance is to assess the variability that would be encountered in some different sampling context or the variance that should be attached to a nonstandard type of comparison. This can be called in general *synthesis of variance*. The simplest example is the estimation of the variance of a mean if n_{S_1} observations were to be made on each of n_{J_1} individuals. This is $(\tau_\epsilon + n_{S_1}\tau_\xi)/(n_{J_1}n_{S_1})$ which can be estimated together with its effective degrees of freedom. Specifically, we may be interested to know how much better will be the precision taking three or four observations on each individual rather than one. If individuals in a study are very different, then there is little benefit from additional replication. But if τ_ϵ is large relative to τ_ξ there will be an $n_{S_1}n_{J_1}$ effect; i.e., if

$$\tau_\xi' = \frac{\mathrm{MS}_\xi - \mathrm{MS}_\epsilon}{n_S} \tag{2.17}$$

is estimated using n_S replicates, then the estimated synthesized variance we obtain changing the number of replicates from n_S to n_{S_1} is

$$\mathrm{var}(\bar{Y}_{..}') \approx \frac{1}{n_{S_1}n_{J_1}}(\mathrm{MS}_\epsilon + n_{S_1}\tau_\xi'). \tag{2.18}$$

Calculations of this sort are important in the design of systems to find a balance between the number of individuals or components that need to be studied and the appropriate number of replicates within each individual.

Example 2.3. More on IPPPSH. Suppose a new clinical trial in heart disease is planned for a population whose pattern of variation in blood pressure is essentially unchanged from the IPPPSH. It is then relevant to ask what would be the predicted standard error of the overall mean blood pressure for a given patient for different numbers of clinic visits and 1, 2 or 3 replicates per visit? We may wish to know by how much the precision is altered, noting that for the IPPPSH the between visit within patient component of variation is the dominant source of variation.

The development outlined above immediately generalizes to hierarchical models with three or more nested effects. For example, the estimated synthesized variance for the overall mean $\bar{Y}_{...}$ obtained from changing the number of patients from n_J to n_{J_1}, the number of visits for each patient from n_K to n_{K_1} and the number of replicates within patient visits from n_S to n_{S_1} is obtained in general as

$$\text{var}(\bar{Y}'_{...}) \approx \frac{1}{n_{S_1} n_{K_1} n_{J_1}} (\text{MS}_\epsilon + n_{S_1} \tau'_\eta + n_{S_1} n_{K_1} \tau'_\xi). \tag{2.19}$$

where τ'_η, τ'_ξ are the estimates of the components of variance based on the observed mean squares and the original degrees of freedom.

We know that each patient mean $\bar{Y}_{j..}$ has a random component $\xi_j + \bar{\eta}_{j.} + \bar{\epsilon}_{j..}$ so that

$$\text{var}(\bar{Y}_{j..}) = \tau_\xi + \frac{\tau_\eta}{n_K} + \frac{\tau_\epsilon}{n_S n_K}. \tag{2.20}$$

Then the estimated synthesized variance for the mean for a given patient j obtained by changing the number of visits and number of replicates within visits as above is

$$\text{var}(\bar{Y}'_{j..}) \approx \frac{1}{n_{S_1} n_{K_1}} (\text{MS}_\epsilon + n_{S_1} \tau'_\eta + n_{S_1} n_{K_1} \tau'_\xi). \tag{2.21}$$

Figure 2.2 shows how the predicted standard error of the overall mean for a patient changes as the number of visits n_{K_1} ranges from 1 to 20 and there are 1, 2 or 3 replicates per visit. Note that when $n_{S_1} = 1$ or $n_{K_1} = 1$. there are only two components of variance. and when both n's are equal to one, then there is only one component of variance for between patients.

There is a dramatic gain in precision in increasing the number of visits from 1 up to 8 or so, after which the precision of the standard error of a given patient mean still improves but by ever decreasing small amounts. Increasing the number of replicated measurements for each visit from 2 to 3, say, has the most noticeable improvement when only a few visits are to be made, but even so the increase in precision is relatively slight compared to the gain in increasing the number of visits per patient.

2.4 Finite population aspects

In the formulation underlying all the above discussion, there are two or more sources of random variability, each represented by independent random variables having distributions defined in principle by properties in a very large number of repetitions. In an alternative essentially equivalent terminology, the resulting values are drawn randomly from a hypothetical infinite population.

Occasionally, especially in some industrial problems, some of the contributions are to be treated as sampled at random from a finite population or may even form the whole of that population. For example, suppose that N_J machines produce nominally identical product and that n_S repeat observations are taken on n_J machines, the machines being selected by random sampling without replacement from the finite population of size N_J. We

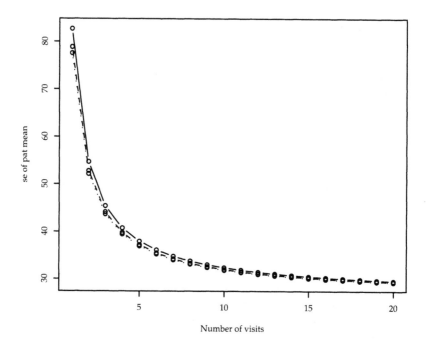

se of pat mean

Number of visits

Figure 2.2 *Estimated standard error of the overall mean for a given patient in a study with the IPPPSH pattern of variation in blood pressure; the standard error is plotted as a function of the number of visits and the number of replicates (1 = solid line, 2 = dashed line, 3 = dotted line) per visit.*

begin as before from the representation

$$Y_{js} = \mu_j + \epsilon_{js}. \qquad (2.22)$$

Suppose that we are interested not in specific comparisons between individual machines but in the contribution of within and between machine variability to overall variability. Subject to the usual assumptions we may estimate $\mathrm{var}(\epsilon)$ from the mean square within machines. If n_J is very small compared with N_J we may proceed as before to treat the μ_j as themselves independent random variables and estimate their variance. But suppose that the N_J machines are the entire population of interest and that n_J is comparable or even equal to N_J. A quadratic measure of variability will be some multiple of

$$\Sigma_{j=1}^{N_J}(\mu_j - \bar{\mu}_{.(\pi)})^2, \qquad (2.23)$$

where $\bar{\mu}_{.(\pi)}$ is the finite population mean.

If there is some *a priori* weighting of the machines the sum of squares can be modified. Note that we no longer express μ_j in terms of a random variable ξ_j because the groups (machines) are regarded as forming a given and fixed population of interest not as representing some hypothetical population of repetitions.

Now if MS_b is the mean square between groups in the standard analysis of variance, replacing what was previously MS_ξ, we have that

$$E(MS_b) = r\Sigma(\mu_j - \bar{\mu}_.)^2 + \sigma_\epsilon^2, \tag{2.24}$$

where $\bar{\mu}_.$ is a mean for the groups actually sampled. If the whole population of groups is observed then $\bar{\mu}_. = \bar{\mu}_{.(\pi)}$.

For random sampling without replacement

$$E\{\Sigma(\mu_j - \mu_.)^2\}/(n_J - 1) = \Sigma_{\alpha=1}^{N_J}(\mu_\alpha - \bar{\mu}_{.(\pi)})^2/(N_J - 1). \tag{2.25}$$

where the expectation on the left refers to the sampling of the finite population. Of course in the important special case when the whole of the finite population is observed the identity is essentially vacuous.

The simplest proof of the general identity is obtained by noting that the expectation on the left, being a symmetric quadratic form in the population means and invariant under constant shifts, must be some multiple of the sum of squares on the right. If then all μ were independent identically distributed random variables of variance τ, say, both sides have expectation proportional to τ, from which the constants of proportionality are as stated. The property set out in (2.25) is sometimes called inheritance from sample to finite population.

It is now very tempting to define the finite population variance by the right-hand side of (2.25), i.e., by a sum of squares of deviations divided by $(N_J - 1)$; then the same estimate is obtained as in the infinite population case studied previously. Temptation is, however, occasionally best avoided. Imagine a universe of individual components formed by mixing the outputs of the N_J machines in equal numbers. Then the variance of a single observation drawn at random from that universe is

$$\Sigma_{\alpha=1}^{N_J}(\mu_\alpha - \bar{\mu}_{.(\pi)})^2/N_J + \sigma_\epsilon^2. \tag{2.26}$$

Thus if we want an operational interpretation as the contribution to overall variability we should define τ_μ to be the finite population sum of squares with divisor N_J. It follows that

$$E(MS_b) = \{n_S N_J/(k-1)\}\tau_\mu + \sigma_\epsilon^2. \tag{2.27}$$

If in particular $n_J = N_J$, so that all the machines in the population are observed, the multiplier of τ_μ is $n_S n_J/(n_J - 1)$, rather than the usual n_S. The most extreme case is $n_J = 2$ when a factor of 2 is involved, as compared with the divisor $(n_J - 1)$.

In theoretical discussions it does simplify some formulae to consider the parameter

$$\Sigma_{\alpha=1}^{N_S}(\mu_\alpha - \bar{\mu}_{.(\pi)})^2/(N_S - 1) = \tau_\mu^*, \qquad (2.28)$$

and we suggest calling this a component of variation.

In the simplest case $n_J = N_S$, and under normal theory assumptions about the ϵ, the distribution of MS_b is proportional to noncentral chi-squared with $n_J - 1$ degrees of freedom. Thus, in particular, formally exact inference for τ_μ/τ_ϵ is based on a noncentral F distribution with $(n_J - 1, n_J(r-1))$ degrees of freedom rather than on the central F distribution. In practice, were an assessment of estimation error in either τ_μ or in τ_μ/τ_ϵ required, it is simplest to approximate to the noncentral chi-squared random variable by a multiple of a central chi-squared random variable, choosing the degrees of freedom to match means and variances.

Example 2.4. Woollen carding (Example 1.5 continued). In the industrial example sketched in Example 1.5, there are four bobbins and 25 ends across each bobbin. These constitute the entire population of bobbins and ends for that machine and in assessing the contribution to overall product variability they must be treated as such. That is, the contribution of a component of variance τ_b for bobbins to the between bobbin mean square is

$$3 \times 25 \times 4\tau_b/3$$

rather than $75\tau_b$ had the bobbins been a sample from a very large population, real or hypothetical, of bobbins. It might appear that a contribution should come from the interaction of bobbins with variation across the card but these are presumed systematic effects and are in essence defined to sum to zero across each bobbin. This leaves the contribution from between Bobbins × Times which is regarded as random leading to an expected mean square for bobbins of

$$\tau_0 + 25\tau_{b \times t} + 100\tau_b,$$

where $\tau_{b \times t}$ is the component of variation for interaction and τ_0 the residual variance.

2.5 Formulation

We now formulate some general principles for the analysis of a broad family of problems, repeating and summarising much of the previous discussion. Suppose that the data can be organized via a number of factors, each characterized by a series of levels and that individual responses are observed in a configuration characterized by cross-classification and nesting in the way already illustrated. In a simply balanced structure there are an equal number of individuals in each cell of the system. At the lowest level of the system, and sometimes other levels also, there will often be nested replication within a factorial cell.

The factors can be of various kinds, for example treatments or quasi-treatments which it is desired to compare, or intrinsic properties of the individuals under study, for example sex, or features of the observational process (observers, sets of apparatus, etc.), or they may correspond to non-specific features, such as blocks in a randomized block experiment or upper-stage units in a sample survey. These distinctions are crucial in design and some aspects of interpretation but rather less so in the present discussion.

We may characterize the data structure by an analysis of variance table showing the main effects and interactions obtained from cross-classified factors and from systems of nesting within groups of individuals. This table identifies the structure independently of actual data and within the framework of ideas to be discussed here suggests a formulation for analysis and interpretation.

We can decompose any arbitrary set of values linearly into terms that correspond to the various terms in the analysis of variance table. main effects, interactions of various orders and residual terms corresponding to any nesting. This is a purely arithmetical, or in a sense geometrical. decomposition with no necessary probabilistic meaning.

We deal, however, with situations where it is fruitful to represent the data by random variables with a distributional structure closely related to the decomposition just described. In this a number of choices have to be made on subject-matter grounds.

A central question concerns how to represent the levels of a given factor. There are a number of possibilities to be distinguished on subject-matter not formal statistical grounds. These are as

(i) an individual main-effect parameter leading to an interpretation based typically on contrasts between levels. This will usually apply when the factor represents a treatment or equivalent. Sometimes, as when the factor represents effects included for error control, the interpretation, while conceptually possible, is of no immediate interest;

(ii) an individual main-effect parameter of interest only as defining a contribution to total variability across the levels in the study, or across some associated finite population. This is illustrated in Example 2.4: this situation arises quite rarely in applications;

(iii) a random variable representing that the main effect is unstructured random variation. One interpretation is that the levels used are effectively a random sample from a hypothetical population of levels. In the most basic situation the random variables associated with different levels of a factor are independent and identically distributed random variables of zero mean and with a distribution characteristic of the factor. In particular. they are described by their component of variance;

(iv) interactions between factors represented by individual parameters are also represented by individual parameters, although, except in (ii), interpretation of, say, any important interaction between two factors is based on the array of means, not on the individual interaction parameters. The latter are to be regarded as formal tools not as direct steps to interpretation, with the exception of (ii);

(v) interactions between factors, at least one of which has a main effect represented by random variables, to be represented by independent random variables of zero mean with a distribution characteristic of the interaction. When one main effect is that of a nonspecific factor, such as blocks or replicates in a randomized design, it will be immaterial whether that main effect is represented as random, but interaction with that factor will typically be regarded as random;

(vi) nested variation represented by independent random variables.

There results a representation of the observations as a linear combination of 'fixed' effects and, in general, a number of random variables each of zero mean and each characterized, in the first instance, by a variance. In any particular application modification of the above schema in the light of special considerations may be needed. For example, if the above prescription consists solely of fixed effects, and thus provides no basis for estimating random variation, it will be necessary to add a notional error random term to the representation and usually to assume that certain interaction terms are negligible. In other applications external empirical or theoretical information may be available about one or more of the variances or that certain interaction terms may be omitted.

Note that when 'fixed' effects are involved it may be convenient to use redundant parameterizations and to impose constraints, such as that interaction terms between 'rows' and 'columns' sum to zero in both directions. These are, however, with the exception of the unusual component of variance interpretation in Example 2.4, no more than conventions with no ultimate effect on interpretation. When, however, random terms are involved such constraints are inappropriate and should not be used.

For balanced data the corresponding decomposition of the data vector into orthogonal components, and thus a decomposition of the sum of squares, leads to unbiased estimates of the variance components and to the estimation of the precision of estimated 'fixed' effect parameters, these being based on relevant sample means. If all random variables can be taken as independently normally distributed the decompositions have strong optimality properties arising from sufficiency.

The initial decomposition is linear and hence changed by nonlinear transformation, for example by taking logs or reciprocals of the original data. We return in Section 6.8 to the systematic development of appropriate transformations. Two key assumptions in applying the decomposition are the approximate normality of distribution and, often more importantly, the strong assumptions of homogeneity of structure involved. For example, some parts of the data might have less random variability than others. This could well distort interpretation as well as degrading efficiency if not allowed for. A further possibility in some contexts is of spatial or temporal correlation between different random terms.

The above discussion is for balanced data and, to an appreciable extent, for approximately normally distributed data. The initial ideas, however, are of much broader importance. Thus suppose the data are unbalanced in the sense that some factor combinations are replicated more frequently than others or even that some are missing. The construction of a formal model and the choice as to which effects are to be regarded as random can proceed as before. The method of analysis is, however, now more complicated even for normally distributed random terms; see Chapter 4. Even more generally we may sometimes regard the starting decomposition as applying to, for instance, the log mean of a Poisson-distributed observation or to the logit transform of the probability of 'success' for binary data. The central point is that the structure of an analogous balanced analysis of variance table provides one systematic basis for the choice of terms, systematic and random, to include in a representation of the data whatever specific distributional assumptions might be suitable. The simple direct analysis of sums of squares no longer applies except perhaps as an approximation.

2.6 Some more theory

We now return to the normal theory, taking simple cases to illustrate general principles.

For a simple random sample, i.e., $n_J = 1$, the analysis of variance decompositions are

1. for data

$$Y_s = \bar{Y}_. + (Y_s - \bar{Y}_.); \tag{2.29}$$

2. for sums of squares

$$\Sigma Y_s^2 = \Sigma \bar{Y}_.^2 + \Sigma (Y_s - \bar{Y}_.)^2; \tag{2.30}$$

3. and for degrees of freedom

$$n_S = 1 + (n_S - 1). \tag{2.31}$$

Because of the orthogonality, the two terms in the observational decomposition are independent. This considerably simplifies the *likelihood* function which is the probability, under the model, of what we have observed as a function of unknown parameters. We find algebraically or more commonly numerically the combination of parameter values that maximize the likelihood. Usually, these are the 'best' estimates with well-known optimality properties; maximum likelihood provides a powerful, general probabilistic procedure for finding estimates of parameters in complicated models.

For the simple random sample, the likelihood contributions factorize. The first term is that of the normal distribution of $\bar{Y}$ around μ, lying in one dimension, and the second lies in $n_S - 1$ dimensions. It is thus essentially that of $n_S - 1$ independent normal variables of zero mean and variance τ_ϵ; this follows from the geometry of least squares or equivalently after orthogonal transformation. Thus its contribution to the likelihood is, except for a constant,

$$\tau_\epsilon^{-n_S/2}\exp\{-\Sigma(Y_s - \bar{Y}.)^2/(2\tau_\epsilon)\}. \tag{2.32}$$

A similar decomposition holds for all balanced arrangements with some random terms. The previous statements about sufficiency and distributional properties follow immediately.

There is a further point important in dealing with unbalanced data. We could build up the full likelihood in the way sketched above and then apply the method of maximum likelihood. In the above example it would lead to the standard estimate of variance dividing by n_S and not $n_S - 1$. In more complex cases the resulting estimates may be entirely unsatisfactory. Note however that if we argued that virtually all the information about τ_ϵ is contained in the second factor and concentrated on that for the estimation we would recover the 'right' divisor $n_S - 1$.

This is connected with general issues of statistical theory about marginal and conditional likelihoods and modifications thereof but in the present context it is called REML (residual, reduced or restricted maximum likelihood), the preferred basis for the formal analysis of unbalanced problems which we address in detail in Chapter 4.

2.7 Bibliographic notes

The key references given in Bibliographic notes for Chapter 1 cover much of the introductory theory here. For a careful theoretical account of analysis of variance, see Scheffé (1959) and, at a more applied level, Snedecor and Cochran (1967). Also many of the books on design of experiments devote substantial space to analysis of variance.

For an excellent introductory account of maximum likelihood and the analysis of variance, see Azzalini (1996).

The importance of distinguishing finite and infinite case in defining components of variance was stressed by Daniels (1939). The extension to sampling a finite population was discussed in detail by Tukey (1950). The device of inherited properties is due to R.A. Fisher. See particularly Irwin

and Kendall (1943–45) and Tukey (1950). The treatment of Example 2.4
in Section 2.4 follows closely Daniels (1939).

2.8 Computational/software notes

The current software for estimating components of variance tends in one of
two main directions. The first is when the components of variance themselves are of primary interest and the fixed effects parameters are regarded
as nuisance parameters. This is the classical field of variance components
and most systems including S-PLUS (Insightful Corporation). R (Ihaka
and Gentleman 1996), GenStat, SAS, SPSS, Stata (StataCorp, 2001). have
something in this style for linear models: see. for example, raov. varcomp
and lme implemented in S-PLUS and R.

Most of the more recently developed software is, however, in the second
direction which is towards estimating fixed effect parameters of primary
interest allowing for random effects with components of variance and covariance. We discuss these issues and appropriate software in Chapters 4
and 5.

Widely used higher-level statistical packages such as R, S-PLUS. Stata.
and SAS are available for most computational platforms now. For example.
R and Stata are available for Unix, Linux, DOS and Macintosh and other
systems. There are active mailing lists for several of these systems which can
often be located via a web search engine such as http://www.google.com.

We are grateful to Dr. W.N. Venables for helpful early advice on available
software for estimating components of variance.

The IPPPSH blood pressure dataset analysed in Example 2.2 is available
from the second author's home page http://www.maths.adelaide.edu.
au/people/psolomon.

2.9 Further results and exercises

1. What is the effect on the estimation of components of variance of
   (a) one outlying observation. (b) one outlying individual?

2. If the primary objective in Situation 3 for between and within variation
   with replication is the estimation of μ how would n_S be chosen? Remem-
   ber that repeat observations on the same individual are likely to cost
   less than the same total number of observations on distinct individuals.

3. A number of similar studies are done with different values of n_S. How
   would one examine the stability across studies of (a) τ_ϵ and (b) τ_ξ?

   It is suggested that τ_ξ may be related to a whole-study characteristic x.
   How would one proceed?

4. Study briefly the resulting effective degrees of freedom for MS_ξ and MS_ϵ

if both components of error are nonnormal. What other aspect needs to be considered in assessing nonnormality?

5. In an investigation of the variation of blood pressure, observations of diastolic and systolic blood pressure were made on 85 patients as follows: on each of three visits, about two weeks apart, duplicate measurements were made under standardized conditions, the duplicate measurements being about one minute apart. For diastolic blood pressure the following sums of squares (mmHg^2) in an analysis of variance table were obtained:

Between patients: $15,310$
Between visits within patients: $3,010$
Between duplicates within visits within patients: 638

Discuss the following points

(a) What assumptions are being implicitly made in the above analysis of variance?

(b) What other analyses might have been set up?

(c) Estimate the components of variance.

(d) What is the estimated standard error of the overall mean blood pressure?

(e) In a new set of data only one observation is made at each visit and there are only two visits per patient. For 100 new patients, predict the analysis of variance table you would expect to see assuming the pattern of blood pressure variation is essentially unchanged. Predict also the estimated standard error of the mean.

6. Obtain the likelihood function corresponding to the model $Y_{js} = \mu + \xi_j + \epsilon_{js}$ when the random components are independently normally distributed. Do this

(a) by noting that the full random vector has a multivariate normal distribution, by evaluating its covariance matrix and then inverting that matrix to obtain the joint density; and

(b) by first finding the likelihood conditionally on the ξ_j and then unconditioning.

Hence show that the minimal sufficient statistic is the sample mean and the two sums of squares MS_ξ and MS_ϵ. Note that these three random variables are independent and give their distributions, one normal and the other two proportional to chi-squared forms.

Write down the likelihood that would result if the only observations were the two sums of squares and show that the resulting maximum likelihood estimates are the standard least-squares-based ones.

7. Express the likelihood in the previous problem in canonical exponential family form. Comment on the relation between this and the structure of the particular parametric combinations for which formally exact inference is available.

8. A balanced configuration tests n_T new treatments each being examined by n_J observers on n_K sets of apparatus with n_S independent observations nested within each of the cells of the three-way $n_T \times n_J \times n_K$ classification. Set out the associated analysis of variance table. Discuss the advantages and limitations of representing all the effects in the resulting analysis of variance, with the exception of the main effect of treatments, by random variables.

Set out the model resulting from the last-mentioned representation and show that the expected values of the mean squares for Treatments, Treatments $\times$ Apparatus and Treatment $\times$ Observers are, respectively,

$$\tau_\epsilon + n_S \tau_{\text{TOA}} + n_S n_J \tau_{\text{TA}} + n_S n_K \tau_{\text{TO}}$$
$$+ n_S n_J n_K \Sigma_i (\mu_i - \bar{\mu}_.)^2 / (n_T - 1);$$
$$\tau_\epsilon + n_S \tau_{\text{TOA}} + n_S n_J \tau_{\text{TA}};$$
$$\tau_\epsilon + n_S \tau_{\text{TOA}} + n_S n_K \tau_{\text{TO}}.$$

Here, for example, τ_{TA} is the variance of the random variables attached to the Treatments $\times$ Apparatus effect.

The key to simple evaluation of such formulae is to note that on substituting the representation of the observations as random variables into the expressions for the sums of squares the cross-product terms all vanish so that the contribution of each source of variability can be evaluated as if it were the only term.

Show that the significance of the treatment effect and, more importantly, the standard error of a treatment contrast cannot be evaluated directly from one line in the analysis of variance table but that a combination of Treatment $\times$ Apparatus, Treatment $\times$ Observers and Treatment $\times$ Apparatus $\times$ Observers can be used to synthesize a suitable error term. What are the effective degrees of freedom of the resulting estimated variance?

CHAPTER 3

More general balanced arrangements

Preamble

In this chapter a number of generalizations of the core issues summarised in the previous chapters are given starting with the generalization to several variables, i.e., to components of covariance and, particularly importantly. components of regression (Section 3.2). An example involving diastolic and systolic blood pressure is discussed in some detail. The special considerations involved when time is a factor are outlined in Section 3.3. The role of and relation to empirical Bayes methods is developed with special reference to microarray data (Section 3.4). The role of components of variance in studying the effect in regression analysis of measurement error in the explanatory variables is described with particular stress on describing how that error is generated (Section 3.5). Some key ideas connected with the design of studies to estimate components of variance are developed (Section 3.7).

3.1 Preliminaries

The previous two chapters cover the key elements in the formulation and analysis of models for balanced data and involving more than one source of random error. In a sense the remainder of the book is concerned with developments of those ideas for more complicated situations and distributional forms. In the present chapter. however, we deal with some specialized points connected with the formulations already described.

3.2 Components of covariance and regression

3.2.1 Introductory remarks and notation

Suppose to begin with that we have one of the situations studied in the previous chapters, but that instead of a single response variable. Y_{js}. the simplest situation, we have a vector of different types of measurement for each (j, s) combination. There is now a notational problem. On the whole. random vectors are most conveniently regarded, essentially by convention. as column vectors. If in the previous formulation the observed random variables are regarded as a $n_J n_S \times 1$ vector each component is now replaced by a set of p components. For some purposes in the present context it is best to regard each Y_{js} as now a $1 \times p$ row vector, so that we have a $n_J n_S \times p$ data matrix Y.

We can now use precisely the previous notation, writing in particular

$$Y_{js} = \mu + \xi_j + \epsilon_{js}. \tag{3.1}$$

All members of this equation are $1 \times p$ row vectors and ξ_j, ϵ_{js} are zero-mean uncorrelated row vectors described by covariance matrices τ_ξ, τ_ϵ. Thus

$$\tau_\xi = E(\xi_j^T \xi_j), \quad \tau_\epsilon = E(\epsilon_{js}^T \epsilon_{js}), \tag{3.2}$$

where the diagonal elements of the τ's are components of variance for the different variables making up Y and the off-diagonal elements are components of covariance. This is equivalent to a particular specification of the covariance matrix of the full $pn_J n_S \times 1$ vector formed by stacking the transposed rows on top of one another, but we shall not use that specification.

A similar argument extends any of the models of Chapters 1 and 2 considered there for a single response variable into a model for a row vector of responses, each component of variance becoming a component covariance matrix. Thus many of the formal considerations extend immediately to vector responses but there are, of course, substantial new issues of interpretation.

Note in particular that any analysis of variance table decomposing the sum of squares of a single variable generalizes to include not only the sums of squares of the separate variables but also sums of products of different variables. When p is reasonably small the result can be written in an analysis of covariance table, as now shown for the special case $p = 2$.

In such a special case it is convenient to use a different notation, denoting the response vector by (U, V), where U and V might represent, for instance, diastolic and systolic blood pressures, or CD4 cell counts and viral load in HIV/AIDS patients. Initially at least we treat the components on an equal footing, i.e., do not regard one as a response to the other considered as explanatory.

The simplest model is in this new notation

$$U_{js} = \mu_U + \xi_j^U + \epsilon_{js}^U, \tag{3.3}$$
$$V_{js} = \mu_V + \xi_j^V + \epsilon_{js}^V, \tag{3.4}$$

for which there are four variances $\tau_\xi^U, \tau_\xi^V, \tau_\epsilon^U$ and τ_ϵ^V as well as covariances

$$\tau_\xi^{UV} = \mathrm{cov}(\xi_j^U, \xi_j^V), \quad \tau_\epsilon^{UV} = \mathrm{cov}(\epsilon_{js}^U, \epsilon_{js}^V). \tag{3.5}$$

The corresponding sums of squares and cross-products take the forms set out in Table 3.1.

Note from the table that the components of covariance split the variability analogously to the split of the components of variance for the univariate responses.

Table 3.1 *Sums of squares and cross-products for a bivariate component of variance model. B: between; W: within.*

	(U,U)	(U,V)	(V,V)
B	$\sum(\bar{U}_{j\cdot} - \bar{U}_{\cdot\cdot})^2$	$\sum(\bar{U}_{j\cdot} - \bar{U}_{\cdot\cdot})(\bar{V}_{j\cdot} - \bar{V}_{\cdot\cdot})$	$\sum(\bar{V}_{j\cdot} - \bar{V}_{\cdot\cdot})^2$
W	$\sum(U_{js} - \bar{U}_{j\cdot})^2$	$\sum(U_{js} - \bar{U}_{j\cdot})(V_{js} - \bar{V}_{j\cdot})$	$\sum(V_{js} - \bar{V}_{j\cdot})^2$

Recall that the covariance,

$$\operatorname{cov}(U, V) = E\{(U - \mu_U)(V - \mu_V)\} \tag{3.6}$$

or its dimensionless equivalent. is a way of capturing the linear part of the relationship between two variables in one number. Covariance is appropriate if the relationship is essentially linear, moderately appropriate if the relationship is monotonic, but inappropriate if the relationship is appreciably nonlinear. If U, V are independent then $\operatorname{cov}(U, V) = 0$.

One way of computing sums of products directly by hand, although of course this is rarely necessary. is to note that

$$4UV = (U + V)^2 - (U - V)^2. \tag{3.7}$$

so that analysis of the sums of squares of the sums and differences of the original variables will determine the sum of products. The probabilistic version of this remark is that

$$4\operatorname{cov}(U, V) = \operatorname{var}(U + V) - \operatorname{var}(U - V). \tag{3.8}$$

3.2.2 Regression interpretations

.We return to the one-way analysis of a $1 \times p$ vector of variables in the form

$$Y_{js} = \mu + \xi_j + \epsilon_{js}. \tag{3.9}$$

where $\tau_\xi = E(\xi_j^T \xi_j), \tau_\epsilon = E(\epsilon_{js}^T \epsilon_{js})$ are $p \times p$ covariance matrices for the between and within group components and μ is a $1 \times p$ vector mean.

It follows directly from the previous results that unbiased estimates of the two covariance matrices are obtained from

$$(\mathrm{MS}_\xi - \mathrm{MS}_\epsilon)/n_S. \ \mathrm{MS}_\epsilon \tag{3.10}$$

where now, for example, MS_ξ is the between-group matrix of mean squares and products. One direct and very general proof of the unbiasedness stems from the identity (3.7) relating sums of products to sums of squares.

We can view the bivariate and more generally the multivariate decomposition in two rather different ways.

If the two components, U and V. say. are treated on an equal footing. we have two covariance matrices for the interpretation of associations at the two different levels. Sometimes it may be helpful to estimate separately the two correlations

$$\text{corr}(\xi_j^U, \xi_j^V), \ \text{corr}(\epsilon_j^U, \epsilon_j^V). \qquad (3.11)$$

In general the correlation or covariance matrices can be interpreted, at least informally, by any of the usual methods for handling such matrices. Just one of the possibilities is to examine the correlation matrices, or their inverses, essentially equivalent to partial correlation matrices, for special independence structures.

The second possibility is that V, say, should be considered as explanatory to the response U. More generally, we may partition the vector Y into two components, interest focusing on the regression of Y_1 on Y_2, defined by matrices of regression coefficients and residual covariance matrices separately at ξ and at ϵ levels. With more than two components the possibility of graphical chain block structures, level by level, might be considered. These are generalizations of Sewall Wright's path analysis in which the variables are represented by nodes, arranged in blocks, and dependencies are represented by edges between pairs of nodes.

For example, at ξ level the matrix of regression coefficients is

$$\tau_{\xi,12} \tau_{\xi,22}^{-1} \qquad (3.12)$$

and the residual covariance matrix is

$$\tau_{\xi,11.2} = \tau_{\xi,11} - \tau_{\xi,12} \tau_{\xi,22}^{-1} \tau_{\xi,21}. \qquad (3.13)$$

Here the suffices 1 and 2 refer to the partitioning of τ_ξ by the components of Y_1 and Y_2.

Thus with two components there are two regression coefficients of U on V, namely

$$\beta_{\xi,UV}, \ \beta_{\epsilon,UV}, \qquad (3.14)$$

regression coefficients from the between and within group structure, respectively. It is clear that there is no necessary connection between these two regression coefficients, although if both represent physically or biologically related sources of variation it will be reasonable to assume that the two regression coefficients are of roughly similar size, and very likely to have the same sign. If, however, the lower level of variation corresponding to the random vector ϵ represents independent measurement error, independent for the two components, then $\beta_{\epsilon,UV} = 0$.

For estimation we compute the matrices of mean sums of squares and products between and within groups and use the relation already noted that for the balanced case

$$E(\text{MS}_\xi) = n_S \tau_\xi + \tau_\epsilon, \qquad (3.15)$$
$$E(\text{MS}_\epsilon) = \tau_\epsilon. \qquad (3.16)$$

If the results with n_J groups and n_S observations per group were merged. i.e., the grouping ignored, then provided n_J and n_S are not too small formally the regression coefficient of U on V estimates

$$(\tau_{\xi,UV} + \tau_{\epsilon,UV})/(\tau_{\xi,VV} + \tau_{\epsilon,VV}). \tag{3.17}$$

In the special case where $\tau_{\epsilon,UV} = 0$ the formula shows the attenuation of the regression towards zero following what is essentially measurement error in V.

We clarify the differences in interpretation of the two regression coefficients in the context of a particular application.

Illustration. Suppose that on a large sample of subjects of stable health and in a narrow age range, measurements are made of blood pressure. U. and Na intake, V. For each subject the observations are repeated some months later. If we ignore possible time trends, we may consider a one-way analysis, between subjects and within subjects.

Even in this simple situation there are complications. Adjustment for country of origin, socio-economic status and possibly general health status may have to be made. Also there is the possibility that some but not all of the variation of V between times within subjects is pure measurement error. Some account of this would be desirable before giving a substantive interpretation to aspects of the components ϵ.

The regression coefficient $\beta_{\epsilon,UV}$ measures the mean increase in blood pressure U when the Na intake. V. of a particular subject varies by one unit, for example 10 mg per day. By contrast $\beta_{\xi,UV}$ is the average difference in the mean blood pressure of two different subjects whose long-run mean Na intakes differs by 10 mg per day.

Now the natural causal interpretation of either regression coefficient is at best tentative. Such an interpretation would imply that individuals increasing their Na intake by 10 mg per day would on the average have an increase in blood pressure determined by the regression coefficient. Such an interpretation of $\beta_{\epsilon,UV}$ would imply such a change on a day-to-day basis as determined from the observed pattern of variation between days within subjects.

The naive interpretation of $\beta_{\xi,UV}$ would imply that if subjects changed their long-run mean Na intake by 10 mg per day then there would be a corresponding change in long-run mean blood pressure as determined by $\beta_{\xi,UV}$. In an observational study. as contrasted with an experiment in which Na intake is randomly allocated to subjects. it is clear that both interpretations involve substantial assumptions and are quite speculative.

Such a direct interpretation of $\beta_{\epsilon,UV}$ would require that all properties that might be explanatory to both U and V are implicitly controlled by inclusion as explanatory variables in the regression analysis or are unimportant. For example. suppose that the amount of physical exercise taken on a day influences both Na intake and blood pressure. Then if the amount of exercise is included as an explanatory variable in the regression equation the regression coefficient on Na intake refers to the effect of changing Na intake without a change in the amount of exercise and, at least from a biological perspective, this is closer to a causal assessment of the effect of Na intake. On the other hand if the amount of exercise is not included in the regression the change in blood pressure anticipated from the regression would be realized on changing Na intake only if the amount of exercise changed appropriately. Note in particular the possibility that it is only exercise that influences blood pressure and that Na intake on its own is irrelevant. We stress that this discussion is purely for illustrative purposes and not based on empirical data.

A somewhat analogous discussion applies to $\beta_{\xi,UV}$. If individuals had been randomized to Na levels the interpretation of the regression coefficient would be unambiguous. In the absence of randomization there may be explanatory variables, observed or unobserved, long-run features of individuals, that are themselves explanatory to both U and V and their effect on the regression coefficient is broadly as before. The interpretation of $\beta_{\xi,UV}$ as if it gave the response to short-term variations of Na intake, i.e., as if it had the same interpretation as $\beta_{\epsilon,UV}$, would require substantial additional assumptions about, for example, the time scale on which the biological phenomenon involved operates.

The argument can be extended to more complicated structures. Suppose, for example, that subjects are grouped by country so that the decomposition of variability is into between countries, between subjects within countries and between times within individuals. There will be corresponding random variables η, ξ, ϵ in the usual representation of the data. The interpretation of $\beta_{\eta,UV}$, the country-level regression coefficient, as applying to either long-run or short-term changes for individual subjects is subject to major uncertainties. Difficulties arising from applying aggregate-level conclusions to individuals are known as ecological.

Example 3.1. The IPPPSH continued. We return now to Example 2.2 and the analysis of blood pressure data from the International Prospective Primary Prevention Study in Hypertension. In Chapter 2 we fitted a nested model to replicated quarterly diastolic blood pressure measurements on 25 men from the four-year cohort. The IPPPSH was a large-scale intervention trial in which diastolic blood pressure was subject to a target control level of < 90 mmHg but in which systolic pressure was not targetted. Figure 2.1 showed the individual patient trajectories for both diastolic and systolic pressures, and it is reasonable to anticipate from the plots that there are appreciable components of covariance between visits within patients, and surely between patients as well.

A simple but natural model for these data is the bivariate nested model extending (3.3) and (3.4) to include an additional random term representing a third level in the hierarchy. The response variables are treated on an equal footing and components of covariance between patients, between visits within patients, and between replicates within patient visits are incorporated into the bivariate model. That is, we assume

$$U_{jks} = \mu_U + \xi_j^U + \eta_{jk}^U + \epsilon_{jks}^U, \tag{3.18}$$

$$V_{jks} = \mu_V + \xi_j^V + \eta_{jk}^V + \epsilon_{jks}^V. \tag{3.19}$$

for which there are now six variances τ_ξ^U, τ_ξ^V, τ_η^U, τ_η^V, τ_ϵ^U and τ_ϵ^V as well as covariances

$$\tau_\xi^{UV} = \text{cov}(\xi_j^U, \xi_j^V). \tag{3.20}$$

$$\tau_\eta^{UV} = \text{cov}(\eta_{jk}^U, \eta_{jk}^V), \tag{3.21}$$

$$\tau_\epsilon^{UV} = \text{cov}(\epsilon_{jks}^U, \epsilon_{jks}^V). \tag{3.22}$$

We are, of course, here ignoring the possibility of trends over time or serial correlation.

An important feature of blood pressure data is that they are positively skew and often well described by a log normal distribution. Solomon (1985) established that the relevant log likelihood for the bivariate nested model (3.18) and (3.19) incorporating components of covariance was maximized by taking a square root transformation of the diastolic data and a log transformation of the systolic data. Of course in practice there are substantial arguments for using a common transformation of such similar variables. That study also showed that the inclusion of components of covariance improved the sensitivity of the transformed data log likelihood. In practice, though, it is reasonable to apply a common log transformation for ease of interpretation and this we do here. A transformation theory for components of variance and covariance models is described in Chapter 6.

Table 3.2 sets out the results of fitting the bivariate nested model to the IPPPSH data.

The results in Table 3.2 demonstrate that the log transformation stabilizes the variance. With respect to the relative components of variance, the variability between visits within patients is the largest source of variation in diastolic pressure, whereas the between patient variability dominates in systolic blood pressure by a small margin. It is likely that the effect of the treatment control of diastolic pressure was to reduce the variability between patients with respect to diastolic but not so strongly for systolic blood pressure. A further observation is that the residual variance components for both blood pressures are roughly equal on the log scale and can reasonably be assumed to be measurement error. The between visits within patient components are also similar under log transformation. Figure 6.1

Table 3.2 *Bivariate hierarchical model with three nested effects fitted to duplicate quarterly diastolic and systolic blood pressure measurements on 25 men from the IPPPSH four-year cohort. The maximum likelihood estimates are for components of variance, covariance and correlation on both the original (first line of each cell) and log transformed (second line of each cell, components of variance and covariance $\times 10^4$) scales. BP: between patients; BV: between visits within patients; W: between replicates within patient visits.*

Source	DBP	SBP	Covariance	Correlation
BP	25.55	190.23	54.78	0.79
	29.55	81.34	37.61	0.77
BV	48.22	170.72	68.30	0.75
	55.97	74.43	48.14	0.75
W	7.73	27.52	6.75	0.46
	9.33	11.78	4.68	0.45
Mean	91.7	149.2		
	4.5	5.0		

in Section 6.8 verifies that the mean-variance relationship apparent in the original data is effectively removed by log transformation.

The components of correlation and variance component ratios are remarkably stable under log transformation. To see this, consider the following parameterisation for the correlation coefficients which determine the overall correlation matrix:

$$(\tau_\xi + \tau_\eta)/(\tau_\xi + \tau_\eta + \tau_\epsilon), \ \tau_\xi/(\tau_\xi + \tau_\eta + \tau_\epsilon). \qquad (3.23)$$

Then for systolic blood pressure, for example, the estimated correlations are 0.93 and 0.49 on the original scale and 0.93 and 0.48 on the log scale.

Correlation coefficients are unchanged by linear transformation and, while the log transformation is locally linear, it is surprising that both correlations are so nearly unchanged after transformation. The components of correlation between patients and between visits within patients are quite high, indicating relatively strong associations between the patient and visit means. This suggests that the treatment control of diastolic blood pressure did not substantially alter the biological association between these two variables. Consideration of these features of the data suggests that log DBP + log SBP and log DBP − log SBP are approximately independent, that is, the geometric mean and ratio of DBP and SBP are approximately independent, and it might be postulated that these two statistics describe different features of the patients.

Log transformation enables diastolic and systolic blood pressure measurements to be treated almost like replicate measurements. Thus an important conclusion which can be drawn from this analysis is that diastolic and systolic blood pressures may sometimes be best analysed together on the log scale, and not separately, as is usually the case.

3.3 Time as a factor

It is quite common to have studies in which on each individual repeat observations are made at a number of time points. Sometimes it may be sensible to treat times as nested randomly within individuals but this should only be done when the predominant source of variability is likely to be virtually totally random, e.g., measurement or sampling error. In general there may well be trends with time or, even if there is essentially stationarity in mean, there may be serial correlation between measurements on the same individual at nearby time points. There may also be a 'practice' effect, or other systematic effects such as seasonal variation.

A general trend could be dealt with by treating time as crossed with individuals and concentrating on interaction as an indicator of random variability. Important serial correlation would have to be specifically analysed as such.

Example 3.2. Time trends and serial correlation. In practice, it may be difficult to distinguish serial correlation from time trends in repeated measures data. Especially with very short time series, it is largely a conceptual issue (i.e., not directly resolvable empirically) whether relationships should be regarded as trends or as short sections of serially correlated systems. Alternatively, it may be the case that both features are present in any given situation.

For illustration, we investigate the possibility of lag-one serial correlation, specifically AR1 with geometrically decaying correlations, and time trends in the IPPPSH blood pressure data.

It is reasonable to suppose that many of the IPPPSH patients improved slowly over the study period, i.e., that their blood pressure levels decreased. The reason this is plausible is that all patients received at least one active hypertensive drug (the betablocker oxprenolol) and diastolic blood pressure was subject to a treatment target level of < 90 mmHg, as discussed briefly in Section 2.1. Thus for these data, a trend representation may well be more interpretable than one based on serial correlation. We note though that the IPPPSH patients were under observation for four years, and it is clear from Figure 2.1 that, although many patients did indeed improve over time and that treatment was apparently effective, other patients slowly deteriorated with trends towards elevated blood pressure levels (age is highly likely to be a relevant factor here, especially amongst the older patients); yet, other patients had blood pressure levels that remained relatively stable over time.

In general, if trends are essentially equally likely to be up or down, this would suggest that serial correlation, in particular a first-order autoregression (AR1), provides a more appropriate interpretation. A priori then, either a trend model or a serial correlation model would be suitable for the IPPPSH data. Indeed, in view of our remarks above about patient blood pressure patterns over time, fitting a model with trend plus AR1 would be justified, primarily to obtain a more realistic assessment of the standard error for trend.

The estimated lag-one serial correlation, which we denote by $\tilde{\rho}_1$, for diastolic blood pressure on the original scale is $\tilde{\rho}_1 = 0.11$ and for systolic, $\tilde{\rho}_1 = 0.13$, both of which are really rather small, suggesting that adjacent observations are not strongly dependent. Given that clinic visits were at least three months apart in this study, the smallness of the correlation is entirely plausible. Moreover, we note that there is considerable variability between visits within patients, as well as evidence from Figure 2.1 of substantial nonlinearity in repeated measurements over time.

The (separate) estimated lag-one serial correlations for the log transformed data are virtually unchanged after transformation, being 0.13 for both diastolic and systolic blood pressure, and we comment on this feature of the data below.

The results from fitting the (linear) random intercepts and slopes model, in which the random intercepts represent the patient effects, are set out in Table 3.3. It is clear from the table that the estimated population slopes are typically small with relatively large associated components of standard deviation. Table 3.3 also gives the results from fitting the combined model with trend plus AR1, for which the estimated standard error of the slope parameter is only marginally smaller than for the model for trend alone.

Using likelihood ratio tests, we compared the goodness of fits of the models in Table 3.3 with each other and with the three-level (nested) hierarchical model (2.6). In each case, the hierarchical model provides a statistically superior fit compared to the random intercepts and slopes model, or to the model combining trends and AR1 (all P values < 0.0001). Adding AR1 to the trend model does not significantly improve the goodness of fit, and the model with serial correlation alone provides the poorest fit to the data. There is therefore strong evidence of substantial nonlinear effects over time, which is confirmed by Figure 2.1 and the results of the analyses in Sections 2.1 and 3.2. But there are no consistent patterns in the observed nonlinear behaviour and this, together with the large between visit within patient variation, provides strong justification for fitting hierarchical models to these data. In this case the subject-matter interpretation of the serial correlation is less interesting. Its presence does not much alter the

Table 3.3 *Random slopes and intercepts models fitted to duplicate quarterly diastolic and systolic blood pressure measurements on 25 men from the IPPPSH four-year cohort for the original and log data. Y: response variable; D: diastolic; S: systolic; μ: grand mean; ξ: patient effect; β: slope; se: standard error: $\sqrt{\hat{\tau}_{\xi b}}$: component of correlation between the intercept and slope random variables; $\tilde{\rho}_1$: lag one autocorrelation. Model M1: time trend only; Model M2: time trend plus AR1.*

Y	$\hat{\mu}(\text{se})$	$\sqrt{\hat{\tau}_\xi}$	$b(\text{se})$	$\sqrt{\hat{\tau}_b}$	$\sqrt{\hat{\tau}_{\xi b}}$	$\tilde{\rho}_1$
D						
$M1$	94.4(1.20)	6.92	−0.18(0.076)	0.44	−0.68	
$M2$	94.4(1.18)	6.86	−0.18(0.075)	0.44	−0.68	0.032
$\log D$						
$M1$	4.54(0.013)	0.077	−2.03(0.86)*	4.96*	−0.71	
$M2$	4.54(0.013)	0.076	−2.04(0.86)*	4.89*	−0.70	0.045
S						
$M1$	150.7(3.16)	17.18	−0.11(0.16)	0.91	−0.58	
$M2$	150.7(3.10)	17.00	−0.10(0.15)	0.89	−0.57	0.046
$\log S$						
$M1$	5.01(0.021)	0.11	−0.72(1.03)*	5.90*	−0.59	
$M2$	5.01(0.021)	0.11	−0.72(0.99)*	5.77*	−0.58	0.050

* Denotes value $\times 10^3$.

precision of the estimated trends, but does, however, reduce the precision of the estimated variance components for variation of slope of trend (not shown).

Note that the random intercepts and slopes are moderately negatively correlated, and that the correlation component is virtually unchanged by log transformation. We have already commented above that the lag-one serial correlations are also almost identical on the original and log scales. This observed stability of the estimated correlations is related to the orthogonality of correlation (and other) parameters under arbitrary transformations. and we discuss this and related issues further in Sections 6.8.2 and 6.10.

3.4 Bayesian considerations

3.4.1 Preliminaries

The adjective *Bayesian* is used in statistical discussions in a number of different senses. The common feature is the use at some point of Bayes's theorem, the simple result in probability theory that allows the inversion of order in a conditional probability statement. This is typically applied to find the posterior distribution of some quantity of interest, i.e., the conditional distribution of that quantity given the data. The quantity of interest may be a parameter or a future observation, for example. For such a posterior distribution to be found, it is necessary to specify not only a model for the data but also a prior distribution for the target quantity and for other unknowns, i.e., the distribution these quantities would have in the absence of the data.

In some situations such an analysis can be accommodated largely or entirely within a directly frequency-based interpretation of probability, i.e., based directly on empirical data plus suitable structural assumptions. We then call the analysis empirical Bayesian and no special conceptual issues are raised. A simple example is developed in Section 3.4.2.

The second possibility is that the prior is a summary of external evidence about the quantities of interest as expressed via a view of probability that is not directly frequentist in nature. To be useful in public discussion such a prior must be evidence-based in some sense, for example be a summary of an expert's opinion on the topic, synthesized perhaps from a large body of data not formally organized. Dangerous though it may be to treat expert opinion, i.e., appeal to authority, like data, there are situations where it cannot be reasonably avoided. The choice of prior is often presented as an aspect of personal belief regardless of its evidence base and that aspect we do not address here.

Thirdly the prior may be in some sense flat, taken to represent initial ignorance and a wish to assess the contribution of the data under analysis, so far as feasible on its own. Much recent applied Bayesian work is of this kind. There are, however, major difficulties with such an approach, at least as soon as a flat prior in many dimensions is involved. It is known, however, that in a small number of dimensions a suitable choice of so-called indifference or reference prior will produce procedures with good frequency properties, for example confidence intervals with approximately the desired properties, and that often, but of course not always, the conclusions will not depend critically on the choice of prior. This seems the soundest basis of. for example, the Bayesian analyses summarised below.

Occasionally the word *Bayesian* is used for any analysis involving more than one level of random variation but this seems quite confusing.

3.4.2 An empirical Bayes analysis

We return to the simplest situation in which random variables Y_{js} for $j = 1, \ldots, n_J; s = 1, \ldots, n_S$ are provisionally assumed to have the structure

$$Y_{js} = \mu + \xi_j + \epsilon_{js}.$$

where, as before, the ξ_j and the ϵ_{js} are independently normally distributed with zero mean and with variances, respectively, τ_ξ and τ_ϵ. There are thus three unknown parameters $(\mu, \tau_\xi, \tau_\epsilon)$ denoted collectively by θ. The discussion so far has concentrated on their estimation.

Suppose temporarily, however, that θ is known and that interest shifts to the mean of one of the groups, say the first, namely $\mu + \xi_1$. or to a linear combination of means of the specific groups, namely $\Sigma l_j (\mu + \xi_j)$. In particular, if the linear combination is a contrast, namely such that $\Sigma l_j = 0$. then interest is in $\Sigma l_j \xi_j$.

Now ξ_1, say, is an unobserved random variable which itself partly determines the distribution of the observations. It is therefore appealing. and can be justified formally from various points of view, that information about ξ_1 is best summarised by its conditional distribution given the data. This is derived by Bayes's theorem. For this we depart briefly from our general notation to denote observed values of random variables, Y. say. as y and to write the probability density function of Y evaluated at y by $f_Y(y)$. Then the required conditional density is

$$f_{\Xi_1|Y}(\xi_1 \mid y) \propto f_{Y|\Xi_1}(y. \xi_1) f_{\Xi_1}(\xi_1). \tag{3.24}$$

Here y denotes the full set of observations but, because only the first group of observations depends on ξ_1. the contribution of the other observations does not affect the required conditional density. Thus the required conditional density is proportional to

$$\exp\{-\Sigma(y_{1s} - \mu - \xi_1)^2/(2\tau_\epsilon)\} \exp\{-\xi_1^2/(2\tau_\xi)\} \tag{3.25}$$

and on writing

$$\Sigma(y_{1s} - \mu - \xi_1)^2 = n_S(\bar{y}_{1.} - \mu - \xi_1)^2 + \Sigma(y_{1s} - \bar{y}_{1.})^2 \tag{3.26}$$

we may discard the contribution of the last term because it produces a factor independent of ξ_1.

After combining the two exponential terms and simplifying we have that the required conditional distribution is normal with mean $\tilde{\xi}_1$ and variance $\tilde{\tau}_{\xi_1}$, where

$$\tilde{\xi}_1 = \{\bar{y}_{1.}/(\tau_\epsilon/n_S) + \mu/\tau_\xi\}\{1/(\tau_\epsilon/n_S) + 1/\tau_\xi\}^{-1}. \tag{3.27}$$

$$1/\tilde{\tau}_{\xi_1} = 1/(\tau_\epsilon/n_S) + 1/\tau_\xi. \tag{3.28}$$

This has the form of an optimally weighted mean obtained from combining the information from $\bar{y}_{1\cdot}$ and that from the distribution of ξ_1 around μ. Effectively the sample mean $\bar{y}_{1\cdot}$ is shrunk towards the general mean.

By the same argument the estimate of any contrast is obtained by shrinking the sample contrast towards zero.

It can be shown that if in the originating model the random variables are not normally distributed, then the above estimates are in a sense the best linear estimates. If, however, the distributions are specified and nonnormal then nonlinear estimates are typically more efficient and in particular the notion of uniform shrinkage towards a central point no longer holds in generality. Indeed, viewed as a point predictor of ξ_1 the quantity $\tilde{\xi}_1$ has a property summarised in the term *best linear unbiased predictor* (BLUP).

The above discussion supposes that θ is known. If n_J is large and the components of θ are replaced by the standard estimates, then the resulting estimate of, say, ξ_1 is changed by $O_p(1/\sqrt{n_J})$ as compared with its intrinsic variability which is $O_p(1)$. The coverage properties of induced confidence intervals based on known θ are changed by $O_p(1/n_J)$ and so, unless n_J is quite small, in which case the approach is of doubtful relevance, the effect of estimating θ is likely to be unimportant. A more refined frequentist theory is possible, but cumbersome; simulation could be used if coverage properties were crucial to interpretation.

What in Bayesian theory is usually called the prior distribution is here the specification of the distribution of the upper random variable ξ, i.e., is an intrinsic part of the probability model for data generation and is not a specification of additional information as is typical in a Bayesian context. In this sense the application is better called empirical Bayesian.

Example 3.3. Smoothing variances for microarray data. It is common in cDNA microarray experiments to have only a small number (often between one and four) of replicate observations available for each gene and to have any number up to 20,000 genes (i.e., cDNA of known sequence content, or expressed sequence tags) spotted onto a single array. It is also frequently the case that genes are replicated across, rather than within, arrays, i.e., to have whole-slide replication. The aim of such an experiment may be to conduct a mass 'gene-screen', or so-called 'low-level analysis' to identify a relatively small number of sequences which show differential expression in the two cell populations being compared. Recall from Example 1.3 that two fluorescently labelled samples are hybridized to a single array in cDNA experiments, and that the expression measurement of interest is the relative fluoresence intensity at each spot.

The simultaneous study of many thousands of genes leads to problems with the application of traditional statistical methods. Difficulties arise for example, from multiple testing when attempting to assess the statistical significance and confidence level associated with each gene. Moreover, when only a small number of replicates is available, the individual gene-specific variances are very imprecisely determined. We consider the latter problem here, for the moment ignoring the issue of multiple testing, and illustrate an empirical Bayes method that has been proposed by Lönnstedt and Speed (2002) to deal with the problem.

We base our discussion on data from a recent microarray experiment conducted by B. Reynolds and his Ph.D. supervisors from the Hanson Centre for Cancer Research/Institute of Medical and Veterinary Science and the Women's and Children's Hospital, Adelaide. The experiment was designed and analysed in collaboration with one of us (PJS), G. Glonek and A. Tsykin. The overall aim of the experiment was to identify genes that play an important role in receptor signalling and leukaemogenesis. Two similar mutant leukaemic cell lines in mice were compared: one line (V449E) proliferates into leukaemia and the other (FIΔ) undergoes differentiation to macrophages and neutrophils. It is hypothesized that there is a set of genes induced specifically in response to expression of V449E that results in its leukaemic effects.

A simple time-course experiment was conducted to compare the two mutants at times zero and 24 hours: it was anticipated that measuring changes over time will distinguish genes involved in promoting or blocking differentiation or that suppress or enhance growth as possible genes involved in leukaemia. The complete experiment was a 2×2 factorial design of block size 2 in which each of the six possible direct hybridizations were performed. Dye-swapped replicates were made of the comparisons between the cell lines at times zero and 24 hours: 16.128 genes were spotted on each array. The design has interesting statistical features which will not concern us here but are discussed in detail in Glonek and Solomon (2002).

Reversing the sign on the dye-swapped replicates at 24 hours. we obtain two log ratios, $M = \log(R/G)$. one from each replicate of each gene: M measures the relative intensity of the two fluorescent dyes at each spot. The fluorescent intensities themselves are intended to represent the concentrations of mRNA in the original cell samples. The log ratios were background-adjusted and normalised within and between arrays to remove the red/green dye biases.

The top picture in Figure 3.1 shows empirical evidence that a number of genes are differentially expressed: this is a scatterplot of the dye-swapped replicates for the FIΔ (red in slide 4) versus V449E (green in slide 4) comparison at time 24 hours. For each gene in each experiment. the log ratios, M, are plotted against each other.

Most of the points are grey and indistinguishable from noise. i.e.. they are consistent with not being differentially expressed. Those genes with a large Malahanobis distance (here taken to be > 20) correspond to the dark grey or black points; the black points correspond to genes which are consistent with differential expression. The Mahalanobis distances were calculated

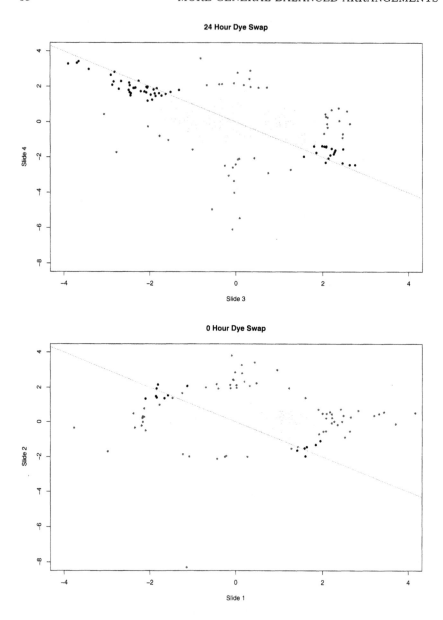

Figure 3.1 *Mahalanobis distance for each gene in the microarray experiment comparing two cells lines at time 24 hours (top picture) and zero hours (bottom picture); the darker points have large Mahalanobis distance (> 20) and the black points correspond to genes consistent with differential expression. For each gene, the log ratio $M = \log(R/G)$ is plotted (R is red and G is green). In slides 1 and 3, the cell line FIΔ was labelled G and the cell line V449E was labelled R; the dye assigment was reversed in Slides 2 and 4.*

using the overall sample mean vector and covariance matrix

$$d_i^2 = (x_i - \bar{x})^T s^{-1}(x_i - \bar{x}), \tag{3.29}$$

where x_i is a two-dimensional vector for the ith gene. The bounds have been calculated to have a very high formal probability of containing the data, in fact 99.995%.

Note that when the dye assignment is reversed in the two cell samples, the differences in the log ratios should be of equal magnitude but opposite sign. At time zero (bottom picture in Figure 3.1) there are almost no points showing any detectable differential expression.

A major challenge is to provide a sound quantitative basis for inferring differential expression. One obvious approach is to select differentially expressed genes on the basis of the mean of their log ratios M, together with the associated standard deviation, and form t-statistics to provide a basis for further analysis. However, very large t's can arise from outlying means which occur relatively frequently in practice, but, more seriously, modestly expressed genes can have tiny standard errors, leading to abnormally large and not necessarily very meaningful t-statistics. Various *ad hoc* solutions have been proposed to deal with this problem, primarily based on setting exclusion cut-offs for genes with small absolute mean values and belonging to the lowest fraction, say 1%, of standard errors, or adjusting the estimated standard errors using tuning constants. Lönnstedt and Speed propose a more formal empirical Bayes approach. The idea is that the information from all genes is combined into estimates of parameters of a prior distribution; these are then combined at the gene level with observed means and standard deviations to form what they call a B-statistic for each gene, which is a Bayes log posterior odds of differential expression. The analysis is similar in principle to, but more complicated than, that outlined above.

The essence of the B-esimator is the following. The log ratios M_{gs} ($g = 1, \ldots, N; s = 1, \ldots, n$) are regarded as random variables from a normal distribution with true mean μ_g and variance τ_g. This assumption is in reasonable agreement with observation. In our study, $N = 16.128$ and $n = 2$. When only a few genes are expected to be differentially expressed in the two cell populations, most genes will have $\mu_g = 0$, and a small proportion will have $\mu_g \neq 0$. Let $I_g = 0$ represent the first case, and $I_g = 1$ the latter case. The parameters (μ_g, τ_g) are treated as independent and identically distributed realisations of random parameters with a specified joint prior distribution. The log posterior odds of differential expression are then

$$B_g = \log \frac{P(I_g = 1|(M_{gs}))}{P(I_g = 0|(M_{gs}))}. \tag{3.30}$$

In particular, the prior distribution of $1/\tau_g$ is assumed to be gamma, and the prior distribution of μ_g given τ_g is taken to be normal; being a conjugate prior, this enables an exact expression to be obtained for the empirical Bayes estimator B. The precise specification is to set, for ν degrees of freedom, $\kappa_g = na/(2\tau_g)$, and to suppose that $\kappa_g \sim G(\nu, 1)$, $\mu_g|\kappa_g = 0$ if $I_g = 0$, and $N(0, cna/(2\kappa_g))$ if $I_g = 1$, for all g. Here, $a > 0$, $c > 0$ are scale parameters, a and ν are hyperparameters in the inverse gamma prior for the variances, and c is a hyperparameter in the normal prior of the nonzero means. Details are given in Lönnstedt and Speed (2002).

Straightforward calculation of the relevant posterior densities shows that, for a gene g,

$$B_g = \log\left(\frac{p}{1-p}\right) \frac{1}{\sqrt{(1+nc)}} \left\{ \frac{a + s_g^2 + M_{g.}^2}{a + s_g^2 + M_{g.}^2/(1+nc)} \right\}^{\nu+n/2}, \quad (3.31)$$

where s_g is the estimated standard deviation (with divisor n rather than $n-1$) for gene g and p is the proportion (unknown) of differentially expressed genes in the experiment, i.e., $p = P(I_g = 1)$ for any g. Clearly increasing $M_{g.}^2$ increases B_g, with a larger effect if the variance is small. If $M_{g.}^2$ is also small, a guarantees that the ratio is not inflated by a tiny variance.

To apply this estimator, we require estimates of the parameters (p, ν, a, c). There are, as yet, no known consistent estimators, so Lönnstedt and Speed recommend fixing p at a sensible value, 0.01 or 0.001 say, then estimating ν, $a|p$, $c|p$, c, a, using all the observed variance estimates to estimate ν and a via the method of moments. The constant c only occurs in the distribution of genes which are differentially expressed, but of course we do not know in advance which ones these are. Usually the whole purpose of a microarray experiment is to identify precisely such genes. However, an estimate of c can be obtained by comparing the observed normal density for a subset of the averages $M_{j.}$, where the subset is the top proportion p of genes with respect to B, with the observed normal density of all the $M_{j.}$.

The top picture in Figure 3.2 plots the B- versus M-statistics for the replicated leukaemic mice experiment at time 24 hours. The shape of the plot is parabolic, with most genes having an average M value near zero. Note that these genes also have the smallest values of B. Higher M's are typically associated with higher B's, but the actual B levels of genes with high averages depend on the variance (thus two genes with the same mean may have quite different B's). A sensible 'cut-off' for differential expression of a gene is unfortunately experiment-dependent and needs to be based on judgement since we do not know p; here, we have taken $p = 0.01$. The V449E cell line is labelled with the red dye; therefore, positive values of the log ratio M correspond to genes which are up-regulated in V449E compared to FIΔ.

With regard to the differentially expressed genes, these lie in both directions, as expected, i.e., some are overexpressed in FIΔ relative to V449E, and vice versa. It is likely that somewhere between 20 to 40 genes are

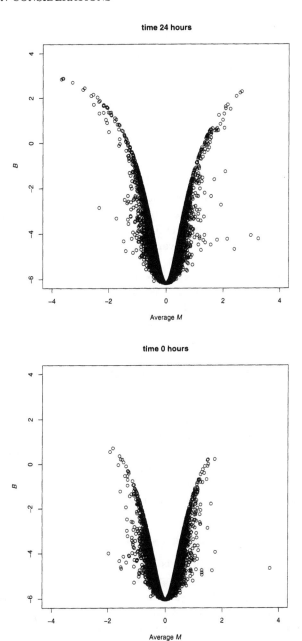

Figure 3.2 *Empirical Bayes estimate B versus the average log ratio M =* $\log(R/G)$ *for each gene in the experiment comparing two cells lines at time 24 hours (top picture) and at time zero hours (bottom picture), ignoring the dye-swap feature of the experiment; FIΔ was labelled G and V449E was labelled R.*

differentially expressed with only a handful of genes (around five) showing evidence of differential expression at time zero hours. Genes with 'extreme' B's defined in this way are worthy of further investigation as candidates for differential expression.

The bottom picture in Figure 3.2 shows the same plot at zero hours, and the comparison of the magnitudes of B demonstrates nicely that there are many more genes differentially expressed at 24 hours than at zero hours.

3.4.3 A fully Bayesian analysis

We now outline a different approach. In the first example of the previous section we took in our initial discussion the parameter $\theta = \mu, \tau_\xi, \tau_\epsilon$ to be known. Later we replaced them by point estimates, arguing that provided n_J is not small errors in estimating θ represent small second-order effects. In a fully Bayesian analysis, however, the prior distribution is over *all* unknowns, including θ, which we may now call a hyperparameter.

Indeed for a Bayesian analysis of the simpler problem in which the focus of interest is on the components of variance and mean, i.e., on θ, a prior for θ is needed. If an evidence-based prior were available, it would be straightforward in principle to form the posterior density of θ as the normalised product of the likelihood and prior and to find the posterior density of any parametric function of interest by integration. There would be no reason for expecting the prior to be of any particularly convenient mathematical form.

If we appeal to the view that flat priors may express lack of knowledge, it can be shown that if the component parameters $\mu, n_S \tau_\xi + \tau_\epsilon, \tau_\epsilon$ are assigned independent flat priors, then the resulting posterior intervals are very close to the frequency-based confidence intervals. Note, however, that from the point of view of expressing external evidence this choice of prior would often be very contrived, depending as it does on n_S, a sample size.

We shall not explore these issues further here.

3.5 Measurement error in regression

The main emphasis in this book is on the estimation of components of variance as parameters representing features of intrinsic interest. Often, however, the real emphasis lies elsewhere and components of variance are of concern only because they affect this primary aspect. One such situation concerns the effect of measurement error in explanatory variables on regression analysis. This issue has a long history and an extensive literature; here we address a few key points.

We suppose that a response variable Y has linear regression on a single explanatory variable X. We treat X itself as a random variable; this is largely for simplicity of exposition. We suppose that for each individual there is a notional true value, X_{tr}, and an observed value, X_{ob}, the latter possibly the average of replicated values. The interpretation of X_{tr} needs consideration in each case. For example it might represent a 20-year exposure to radon gas, measurable in principle if not in practice by continuous dosimetry. Again X_{tr} might represent the mean blood pressure of an individual over a period of stable health as determined by continuous monitoring over waking hours. Here X_{ob} is the observation actually recorded for the individual in question.

The structure of the measurement error $X_{ob} - X_{tr}$ is crucial to the following discussion and can be studied only indirectly and by examination of the nature of the measurement process, X_{tr} itself being unobservable. Biases, especially biases that vary across the range of the measurements, may be a major concern but we shall ignore them here. Also we suppose that, possibly after transformation, the relation between X_{tr} and X_{ob} is stable across the range of the data: for example, if proportional errors are expected a log transformation would be suitable.

Two extreme forms of error relation are the classical and the Berkson error models. In these we have, respectively, that

$$X_{ob} = X_{tr} + \epsilon_{ob.tr}, \tag{3.32}$$

$$X_{tr} = X_{ob} + \epsilon_{tr.ob}. \tag{3.33}$$

Here each error term is uncorrelated with, and sometimes may be taken to be independent of, the variable on the right-hand side.

In terms of components of variance, we have in the two cases that

$$\tau_{X_{ob}} = \tau_{X_{tr}} + \tau_{\epsilon_{ob.tr}}, \tag{3.34}$$

$$\tau_{X_{tr}} = \tau_{X_{ob}} + \tau_{\epsilon_{tr.ob}}. \tag{3.35}$$

This shows the radically different, indeed quite opposing, relations between the corresponding components of variance in the two models. Intermediate forms are possible.

The classical model is usually an appropriate one when the deviation between observed and notional true value is a pure sampling error or due to the use of a convenient rather than a gold standard method of measurement. The Berkson model is less commonly suitable but is relevant in two very different situations. One is experimental in which the X_{ob} are the target values of X and are used in analysis; the unobserved X_{tr} are the realized values of X. They deviate randomly from the target values and determine the response Y. A second situation, relevant to the radon study mentioned above, is where X_{ob} is a heavily smoothed value, for example the radon emission for a region, from which the true value deviates randomly.

Now suppose that the response Y has a linear least-squares regression on X_{ob} in the form

$$Y = \alpha_{tr} + \beta_{tr}X_{tr} + \epsilon_{Y.X_{tr}}, \tag{3.36}$$

where the error is uncorrelated with the variable on the right-hand side. If the regression is a full linear regression rather than merely a linear least-squares regression the error is independent of X_{tr}.

Now in terms of linear least-squares regression we may invert the relation for X_{ob} given X_{tr} into

$$X_{tr} = \mu_X + \beta_{ob}(X_{ob} - \mu_X) + \epsilon_{tr.ob}, \tag{3.37}$$

where $\mu_X = E(X_{tr})$ and the least-squares regression coefficient is calculated as

$$\beta_{ob} = \text{cov}(Y, X_{ob})/\text{var}(X_{ob}) = \lambda\beta_{tr}, \tag{3.38}$$

where the attenuation factor λ is given by

$$\lambda = \tau_{X_{tr}}/\tau_{X_{ob}} = \tau_{X_{tr}}/(\tau_{X_{tr}} + \tau_{\epsilon_{ob.tr}}). \tag{3.39}$$

Also the variance of the error in the regression of Y on X_{ob} is $\lambda\tau_{\epsilon_{ob.tr}}$.

We have related here the linear least-squares regressions of Y on X_{tr} and on X_{ob}. A full linear regression equation requires that the error term is *independent* of the explanatory variable on the right-hand side. By contrast in a linear least-squares regression the error is only uncorrelated with the explanatory variable. Both equations considered here are simultaneously full linear regression systems only rather exceptionally, primarily when all the random variables concerned have normal distributions.

Suppose that interest lies in the regression of Y on the true value X_{tr}. It is defined in this formulation by an intercept, a slope and a component of variance about the regression line. If we observe Y and X_{ob} we can estimate the corresponding parameters in their regression relation and also μ_X and $\tau_{X_{tr}}$. To estimate in particular β_{tr} we thus need one more piece of information, essentially to determine λ. This is usually best obtained via an independent study of the measurement process for determining X, leading to an estimate of $\tau_{\epsilon_{ob.tr}}$. Alternatively a sensitivity analysis can be made over the plausible range of values of $\tau_{\epsilon_{ob.tr}}$.

Thus the general effect of measurement error is a flattening or so-called attenuation of the regression by an amount determined by the relative magnitude of the component of variance due to measurement error in X to that for the real variation in X. Note that error has to be quite appreciable as a component of the total variation of X_{ob} for attenuation to be a really major issue.

The discussion in the Berkson case is much more direct since we can substitute the defining relation

$$X_{tr} = X_{ob} + \epsilon_{tr.ob} \tag{3.40}$$

directly into the regression relation on X_{tr} to obtain

$$Y = \alpha_{tr} + \beta_{tr}X_{tr} + \epsilon_{Y.X_{tr}} + \beta_{tr}\epsilon_{tr.ob}, \tag{3.41}$$

a regression relation of unchanged slope but increased variance about the regression line. Provided information essentially equivalent to knowledge of λ is available it is possible to decompose the observed variance about the regression line into a component present in the underlying relation with X_{tr} and a component induced by the measurement error in X.

In summary in the classical case measurement error flattens the regression, whereas in the Berkson case the effect is to increase the variance about the regression line.

Essentially the same arguments apply to more general cases, including polynomial and nonlinear regression, although the qualitative conclusions need careful specification.

It is convenient for further discussion to regard the classical and Berkson error structures as special cases of the general formulation that

$$X_{ob} = \tilde{X} + \epsilon_{ob}. X_{tr} = \tilde{X} + \epsilon_{tr}, \tag{3.42}$$

the previous special cases being recovered when one or the other ϵ is zero. Thus, to study multiple linear regression, suppose that we have a response Y on a vector of true explanatory variables which we write in the form

$$Y = \mu + \beta_{tr}^T X_{tr} + \epsilon_{Y.tr} \tag{3.43}$$

and that, taking the classical error structure, we have in the linear least-squares sense

$$X_{tr} = \mu + B_{tr.ob} X_{ob} + E_{tr.ob}. \tag{3.44}$$

Here $B_{tr,ob}$ is a matrix of regression coefficients of X_{tr} on X_{ob}, each row of which gives the regression coefficients for a particular component of X_{tr}. In fact

$$B_{tr,ob} = \tilde{\Sigma}(\tilde{\Sigma} + \Sigma_{\epsilon_{ob}})^{-1}, \tag{3.45}$$

where the Σ are the relevant covariance matrices. It follows that for the linear least-squares regression coefficients of Y on X_{ob} we have

$$\beta_{ob}^T = \beta_{tr}^T B_{tr,ob} = \beta_{tr}^T \tilde{\Sigma}(I + \tilde{\Sigma}^{-1}\Sigma_{\epsilon_{ob}})^{-1}\tilde{\Sigma}^{-1}. \tag{3.46}$$

For example, suppose that just one component, say the first, of the explanatory variables is subject to error. Then all elements of $\Sigma_{\epsilon_{ob}}$ except the $(1,1)$ element are zero. Explicit calculation of the above matrix inverse is now possible. If $\tilde{\sigma}^{rs}$ denotes the (r,s) element of $\tilde{\Sigma}^{-1}$, then $\tilde{\sigma}_{1,rem}^2 = 1/\tilde{\sigma}^{11}$ is the conditional variance of $\tilde{X}_1$, the first component of $\tilde{X}$ about its linear regression on the remaining components and $\tilde{\beta}_{1s.rem}$, the linear regression coefficient of $\tilde{X}_1$ on $\tilde{X}_s$, adjusting for all other components, is $-\tilde{\sigma}^{1s}/\tilde{\sigma}^{11}$. It follows after some matrix calculation that

$$\beta_{ob,1} = \beta_{tr,1}\tilde{\sigma}_{1,rem}^2/(\tilde{\sigma}_{1,rem}^2 + \sigma_{ob}^2), \tag{3.47}$$

$$\beta_{ob,j} = \beta_{tr,j} + \beta_{tr,1}\tilde{\beta}_{1s.rem}\tilde{\sigma}_{ob,1}^2/(\tilde{\sigma}_{1,rem}^2 + \sigma_{ob}^2). \tag{3.48}$$

The important qualitative interpretation of this is as follows. The regression coefficient on the component variable subject to error is attenuated exactly as in the case of simple linear regression. The regression on that component is, however, shared out among any other component explanatory variables with which the error-prone component is correlated. Thus the effect of measurement error may be to induce a spurious dependence on well-measured variables with which a badly measured variable is associated.

From the point of view of the present book a key aspect is that the discussion hinges on the relative magnitude of various components of variance and covariance.

3.6 Heterogeneous variability

Almost all the representations considered in this book involve assumptions, sometimes quite strong, that the component of variance for a particular source of variability is constant. There is no difficulty if entirely separate sets of similar data are allowed to have one or more components of variance estimated separately; there may indeed be interest in examining which components of variance are very different and which similar between different sets of data.

Situations in which for a single set of data the same broad source of variability has nonconstant variance are more difficult, however. We now consider briefly the simplest such situation. Let Y_{jk} denote a $n_J \times n_K$ array of observations, i.e., a balanced two-way array without replication. We start with the simplest additive model

$$Y_{jk} = \mu + \alpha_j + \beta_k + \epsilon_{jk}, \tag{3.49}$$

where the ϵ_{jk} are independent random variables of zero mean. Now suppose that $\mathrm{var}(\epsilon_{jk}) = \tau_k$, say, may depend on k. It is now immaterial whether the α_j are regarded as random or as fixed.

There are two possibilities as regards the column effects. If they are random, so are the τ_k because the labelling of the columns is arbitrary. If, however, the columns are individually meaningful and therefore contrasts of the β_k are of potential interest, it will typically be sensible to regard the τ_k likewise. If n_K is large it may be reasonable to take an empirical Bayes approach and to regard the τ_k as random variables, but interest will return to the individual values. Yet another possibility is that τ_k depends on explanatory variables connected with column k or with the associated mean parameter β_k. In some of the latter cases a transformation of the response variable may be helpful.

We concentrate here on the case where the τ_k are of individual interest and regarded as arbitrary parameters. Maximum likelihood estimates can be shown to be inconsistent unless both n_J and n_K are large. An unbiased estimate can be obtained by evaluating and correcting for the bias of the naive estimate proportional to

$$\Sigma_j R_{jk}^2, \tag{3.50}$$

where $R_{jk} = Y_{jk} - \bar{Y}_{j\cdot} - \bar{Y}_{\cdot k} + \bar{Y}_{\cdot\cdot}$.
  This leads to the estimate

$$\tilde{\tau}_k = \frac{n_K}{(n_J - 1)(n_K - 2)} \left\{ \Sigma_j R_{jk}^2 - \frac{\Sigma_{j,l} R_{jl}^2}{n_K(n_K - 1)} \right\}. \tag{3.51}$$

This will be efficient at most locally when all the τ_k are approximately equal. It can be shown that a different method of estimation is needed if $n_K = 2$. More detailed inference can be based on the estimated variance of $\tilde{\tau}_k$, using the device of effective degrees of freedom.

3.7 Design issues

3.7.1 Preliminaries

The main emphasis throughout this book is on the use of components of variance in the analysis and interpretation of data. We now discuss briefly some issues connected with the design of investigations to estimate components of variance. Note that we do not use the term design of *experiments* in that the objective is typically the study of patterns of variation as they exist rather than the assessment of interventions imposed under the control of the investigator.

Nevertheless many of the general principles of experimental design. especially those common with the principles of sampling, apply. In particular. randomization may play an important indeed sometimes crucial role. For example, if one component of variance represents pure measurement error assessed by the variation between duplicate measurements of the same individual, concealment of the identity of duplicates is usually highly desirable to avoid substantial underestimation of the variability. This is frequently best achieved by some element of objective randomization. Again if variation between, say, individual patients is of concern it is in principle desirable that the patients should be chosen from the population of interest by some form of random sampling, for example by stratified or two-stage random sampling. Where this is not possible, the essential dependence of the estimated components on the patients who happen to have been available has to be noted, making the estimates more a summary description of those particular patients than having a deeper interpretation.

In the design issues discussed below cost considerations enter unavoidably. Often, however, these have to be incorporated rather qualitatively. Also there may frequently be unavoidable practical constraints on some features of the design. A final difficulty in the way of any formal treatment of optimal design is that the problem is nonlinear in the sense that the optimal design depends on the unknown parameters under investigation.

3.7.2 Some general considerations for nested structures

We begin with hierarchical, i.e., purely nested, arrangements. In any balanced arrangement the degrees of freedom available increase as one goes deeper into the arrangement. Thus the lower components of variance will be estimated relatively more precisely, and often much more precisely, than the upper components. This is probably if anything the reverse of what is required. Therefore we concentrate on arrangements in which at each level the design splits into at most two parts. That is, in some parts there is just one strand and in others two, the design thus being unbalanced in a relatively simple way.

We start with the simplest situation of n_J groups, n_{J_1} of which have one observation and n_{J_2} of which have two observations split at the lower level to yield each independent observations nested within the group. Under the standard model there are thus four mean squares. From the groups with just one observation per group there is $MS_\xi^{(1)}$ with $n_{J_1} - 1$ degrees of freedom estimating $\tau_\xi + \tau_\epsilon$. From the groups with two observations each there are $MS_\xi^{(2)}$ estimating $2\tau_\xi + \tau_\epsilon$ with $n_{J_2} - 1$ degrees of freedom and $MS_\epsilon^{(2)}$ estimating τ_ϵ with n_{J_2} degrees of freedom. Finally there is a single degree of freedom comparing the means of the two parts giving a mean square $MS_\xi^{(12)}$ with expectation

$$2\tau_\xi(n_{J_1} + n_{J_2})/(n_{J_1} + 2n_{J_2}) + \tau_\epsilon. \tag{3.52}$$

To analyse the data from such a system we may maximize the likelihood derived from the four mean squares; note that this is essentially a version of what in Section 4.6 will be discussed as REML. For the present discussion of design we suppose for simplicity that n_{J_1} and n_{J_2} are both large and ignore the information contained in the single degree of freedom comparing groups. Then the information matrix for $(\tau_\xi, \tau_\epsilon)$ can be found and on inversion shows that asymptotically

$$\text{var}(\hat{\tau}_\xi) = 2(n_{J_1}\gamma_1^2 + n_{J_2}\gamma_2^2 + n_{J_2}\gamma_\epsilon^2)\Delta, \tag{3.53}$$

$$\text{var}(\hat{\tau}_\epsilon) = 2(n_{J_1}\gamma_1^2 + 4n_{J_2}\gamma_2^2)\Delta, \tag{3.54}$$

where

$$\gamma_1^{-1} = \tau_\xi + \tau_\epsilon, \tag{3.55}$$

$$\gamma_2^{-1} = 2\tau_\xi + \tau_\epsilon, \tag{3.56}$$

$$\gamma_\epsilon^{-1} = \tau_\epsilon, \tag{3.57}$$

$$\Delta = (n_{J_1}n_{J_2}\gamma_1^2\gamma_2^2 + n_{J_1}n_{J_2}\gamma_1^2\gamma_\epsilon^2 + 4n_{J_2}^2\gamma_2^2\gamma_\epsilon^2)^{-1}. \tag{3.58}$$

Table 3.4 *Variance of estimate of upper variance component from mixed design with a proportion r of pairs and 1−r of single observations divided by variance for same number of individuals, lower variance component assumed known. The ratio of variance components is λ.*

			r		
λ	0.1	0.25	0.5	0.75	0.9
0.2	4.69	1.92	0.99	0.67	0.57
0.5	3.89	1.72	0.97	0.70	0.60
1	2.85	1.46	0.94	0.74	0.67
2	1.89	1.20	0.93	0.80	0.75
5	1.23	1.03	0.94	0.89	0.86

If now cost considerations are to be drawn into the design we suppose that the unit of cost is that of choosing a new individual and making one observation on it. Suppose that an additional observation on that individual costs c. If interest focuses on the upper variance component, we could aim to minimize (3.53) for a given total cost, i.e., for given $n_{J_1} + n_{J_2}(1 + c)$. There are a number of alternative objectives, including estimation of the ratio τ_ξ/τ_ϵ.

Table 3.4 addresses a special aspect not based on explicit optimization. If the ratio $\lambda = \tau_\xi/\tau_\epsilon$ is large it is plausible that for estimation of τ_ξ it is adequate to have most of the individuals measured just once. Rather more generally if τ_ϵ were known *a priori*, for example from previous data or from theory, we would take $n_J = n_{J_1}$ and $n_{J_2} = 0$ leading to

$$\mathrm{var}_K(\hat{\tau}_\xi) = 2(\tau_\xi + \tau_\epsilon)^2/n_J, \tag{3.59}$$

the variance for a known lower component.

For comparison we therefore fix n_J, write $r = n_{J_2}/n_J$ and examine

$$\mathrm{var}(\hat{\tau}_\xi)/\mathrm{var}_K(\hat{\tau}_\xi) \tag{3.60}$$

as a function of r and of the ratio λ.

As is to be expected, the larger r, i.e., the larger the total number of individual observations, namely $(1 + r)n_J$, the smaller is the resulting var$(\hat{\tau}_\xi)$. At $r = 1/2$ the resulting estimate has a variance slightly smaller than would be achieved by knowing τ_ϵ, and the fractional improvement with increasing r is relatively slow unless λ is small. This suggests that, depending of course on the costs involved, the choice $r = 0.5$ might often be sensible, corresponding to an equal mix of single and paired values.

Similar arguments apply to more complicated nested arrangements. For example with three levels of random variation represented by random variables η, ξ, ϵ suppose that there are m_{ij} individuals split into i sections at ξ

level and into j components at ϵ level for $i, j = 1, 2$. Excluding the three degrees of freedom for comparing the four configurations of individuals there are then eight different mean squares each with expectation a simple function of the three components of variance. It might often be reasonable to restrict attention to $n_{J_{11}} = n_{J_{22}} = 0$, so that all individuals generate just two observations.

3.7.3 Cross-classified structures

The discussion for cross-classified arrangements is similar in principle but of course different in detail. For a two-way arrangement with replication within cells and both rows and columns represented by random variables there will be either three or four components of variance depending on whether interaction is included. As before, if there is substantial interest in the upper components of variance some replication can reasonably be sacrificed at the within-cell level, at the cost of either some complication or some inefficiency in analysis.

We may consider designs with nested observations within cells in which some cells have no observations and some just one observation, whereas others have two, each row of the design and each column of the design preferably having the same proportion of the three types. In some situations it may be important that every cell has at least one observation. Balanced arrangements can be achieved in simple cases by repetition of a $k \times k$ Latin square, assigning some letters of the square to each level of replication. Thus repetition of a 3×3 Latin square would generate designs with equal numbers of 0, 1 and 2 observations per cell, the numbers of rows and columns both being a multiple of 3.

A similar idea applies to a row by column arrangement without replication within cells in which only particular cells have observations. Here the number of replications is zero or one, rather than zero, one or two as previously.

Yet another possibility is that one of the classifications, say the rows, represents a fixed effect.

3.8 Bibliographic notes

The bivariate nested model for the IPPPSH blood pressure data in Section 3.4 follows Solomon (1985). Sy et al. (1997) propose a bivariate stochastic model for analysing and predicting immunological markers for HIV/AIDS; their model incorporated serial correlation via a bivariate stochastic process. J. Taylor and B.D. Ripley assisted with preliminary aspects of the analysis described in Example 3.2.

There is an extensive literature and history of the effect of measurement and sampling error in regression studies. See, for example, Carroll et al.

(1995). The treatment here largely follows Reeves et al. (1998). For an introduction to chain block graphs, see Cox and Wermuth (1996). Robinson (1991) gives a useful overview of BLUPs.

Lönnstedt and Speed (2002) propose the empirical Bayes estimator B and compare its performance to other statistics used to infer differential expression from replicated microarray data; B appears to be marginally superior to basing inference simply on the log ratio intensities, or on modified t-statistics. The method described in Example 3.3 has been extended to the estimation of B-statistics in a linear model by Lönnstedt et al. (2002). Efron et al. (2001) also have studied the use of empirical Bayes methods for the analysis of microarray data.

The account of heterogeneous variances in the two-way arrangement is based on the thorough discussion of Ehrenberg (1950), some of whose arguments anticipate the notions of MINQUE and such-like estimates.

Khuri (2000) should be consulted for a systematic review and bibliography of work on design for the estimation of components of variance. Some of the earlier work establishes the formal optimality of balanced designs under simple constraints. In the 1950s and 1960s there was substantial work by R.L. Anderson and associates starting from Anderson and Bancroft (1952) and reviewed by Anderson (1975). The approximately equal spreading of degrees of freedom across the various levels was studied in particular by Bainbridge (1965). Goldsmith and Gaylor (1970) compared 61 designs for a two-stage nested model. Much of the earlier work used simple, not necessarily efficient, estimates in the unbalanced cases.

3.9 Computational/software notes

At present, neither S-PLUS nor R have routines for directly fitting multivariate nested or crossed models. However, it is possible to fit bivariate and higher-level random effects models using `lme` by specifying an additional factor for the multiple responses together with the appropriate correlation structure via `corStruct`.

ASREML handles multivariate data with random effects in a variety of ways, as does MLwiN. Stata and SAS contain procedures for handling multiple random effects in normal theory linear models.

Within S-PLUS and R, `lme` allows estimation of serial correlation, random slopes, and so on. In Example 3.2, we estimated the first-order serial correlation and fitted random slope models using the `corAR1` option and other appropriate arguments in `lme`.

The empirical Bayes plots in Figure 3.2 were produced using the **sma** library (Statistics for Microarray Analysis, version 0.5.6 by Bolstad et al., 2001) within R, in particular, via the function `stat.bayesian` written by I. Lönnstedt and Y.H. Yang. **sma** can be downloaded from the Comprehensive R Archive Network (CRAN) at `http://cran.r-project.org`.

The **sma** library is currently being expanded and incorporated into software available from the Bioconductor Project http://bioconductor.org. The leukaemic mice data analysed in Example 3.3 can be obtained from The University of Adelaide's Microarray Analysis Group website http://www.maths.adelaide.edu.au/MAG.

Venables and Ripley (1999), Snijders and Bosker (1999) and Pinheiro and Bates (2000) contain important computational work and guidance for fitting random effects and other models. Otherwise, the computational notes made at the end of Chapter 2 are also pertinent here.

3.10 Further results and exercises

1. If a log transformation is desirable what effects would one expect to see in the untransformed data?

2. Consider an abstraction of the situation outlined in Example 1.5. Suppose that a rectangle of constant width is cut in the perpendicular direction into $m(= 25)$ strips in an irregular way and the width of the individual strips measured at a number of cross-sections (times). Note that the widths sum to a constant, namely the width of the rectangle. Show that in the resulting analysis of variance the sum of squares for times is zero and that the interaction terms form a finite population of fixed 'row' sums. Discuss why this example is exceptional in not obeying the commonly valid principles that models with interaction should contain the corresponding main effects and that random interaction terms should not be constrained to sum to zero. For general discussion, see Nelder (1977) and for a very general treatment of the issues involved McCullagh (2000).

3. Suppose that pairs of observations are taken on a group of individuals, that there is some possibility of a time trend and that the time spacings of the observations for different individuals are not all the same (and are known). Outline a possible initial analysis.

4. Discuss the possibilities of an empirical Bayes analysis of the problem of heterogeneous variances (i) assuming an inverse gamma prior distribution for the variances, (ii) assuming a normal distribution for the log variances.

CHAPTER 4

Unbalanced situations

Preamble

Unbalanced data are discussed, starting with the simplest instance of the unbalanced one-way classification. It is argued that in many cases one of two simple forms of analysis is adequate, which one depending on the relative magnitudes of the two components of variance (Section 4.2). The argument is extended to the important situation where the effective observations for analysis are not originating values as recorded by the investigator but rather estimates of parameters derived by fitting models summarising distinct sets of data (Section 4.3). An example about Australian intensive care units is analysed in some depth (Section 4.4). A further application is to the synthesis of studies, i.e., so-called meta-analysis or overview (Section 4.5). An example from genetic epidemiology is described. A more formal analysis uses a modification of maximum likelihood, REML, the basis of which is described (Section 4.6).

4.1 Introduction

The previous discussion has focused on sets of data and analyses that are balanced in a natural sense. In effect the incidence matrix specifying the number of observations in each cell defined by the explanatory factors of concern is very simple, often specifying equal numbers of observations in each cell. This combined with an underlying linearity implies that simple statistics such as row, column, ... means both have a direct interpretation and are the basis of simple analyses leading, in particular, to estimates of components of variance that are efficient, at least under the most natural starting assumptions.

As soon as either the data are unbalanced or the most appropriate model is nonlinear, for example a generalized linear model extension, the formally efficient analysis becomes more complicated. Extensive computation is likely to be involved. While with appropriate software this may not be a major hurdle there is an attendant disadvantage of loss of transparency. That is, the path between the data and the conclusions may be unclear and there is some loss of understanding and sometimes dangers of misinterpretation arising from this, notably by the use of inappropriate models and hence potentially misleading analyses. Therefore, while we discuss efficient methods elsewhere we give here a simpler approximate discussion that covers some at least of the commoner problems, often with negligible loss of efficiency.

4.2 One-way classification

To avoid undue notational complexity, we depart here from our usual notation and start with the simplest situation of observed random variables $Y_{js}(j = 1, \ldots, n_J; s = 1, \ldots, r_j)$, where the replication numbers r_j are in general not all the same. We shall see later that by regarding Y_{js} not as a primary observation but as the outcome of an initial stage of analysis this is a much more general situation than it might at first sight appear. We write, as before, $\bar{Y}_{j.}$ for the group means and explore the model

$$Y_{js} = \mu + \xi_j + \epsilon_{js}, \tag{4.1}$$

where the random variables on the right-hand side are independently normally distributed with zero mean and variances, respectively,

$$\tau_\xi = \sigma_\xi^2, \quad \tau_\epsilon = \sigma_\epsilon^2. \tag{4.2}$$

Now $\bar{Y}_{j.}$ is normally distributed with mean μ and variance $\tau_\xi + \tau_\epsilon/r_j$. If both variance components are known, examination of the likelihood function shows the sufficient statistic for μ, expressed in the form of an unbiased estimate of μ, to be

$$\hat{\mu}_K = \{\Sigma \bar{Y}_{j.}/(\tau_\xi + \tau_\epsilon/r_j)\}\{\Sigma 1/(\tau_\xi + \tau_\epsilon/r_j)\}^{-1} \tag{4.3}$$

with variance

$$1/\Sigma(\tau_\xi + \tau_\epsilon/r_j)^{-1}. \tag{4.4}$$

When the variance components τ_ϵ, τ_ξ are unknown, asymptotically efficient estimates of μ are obtained by replacing the variance components by efficient estimates. In fact, because of the orthogonality of the mean and the variance components, consistency of estimation of the latter would be enough to ensure first-order asymptotic efficiency in estimating μ and the optimal asymptotic variance is (4.4), unchanged by estimation of the components of variance.

While in this simple case it is easy enough to find the maximum likelihood estimate of μ or some asymptotic equivalent thereto, to help tackle more complex cases we explore simple noniterative estimates. These use two extreme forms for the estimate of μ appropriate when the error from a single variance component is predominant. They are, respectively, the overall mean, i.e., the average of the separate group means weighted by sample size, and the unweighted average of the separate means. We write $r_. = \Sigma r_j$. Then

$$\bar{Y}_W = \Sigma_{j,s} Y_{js}/r_. = \Sigma r_j \bar{Y}_{j.}/r_., \tag{4.5}$$

$$\bar{Y}_U \;=\; \Sigma \bar{Y}_{j.}/n_J. \tag{4.6}$$

A direct calculation shows that

$$\text{var}(\bar{Y}_W) \;=\; \tau_\xi \Sigma r_j^2/r^2 + \tau_\epsilon/r., \tag{4.7}$$

$$\text{var}(\bar{Y}_U) \;=\; \tau_\xi/n_J + \tau_\epsilon \Sigma r_j^{-1}/n_J^2. \tag{4.8}$$

These variances can most easily be compared with one another and with the information limit by supposing that the variation in the r_j is relatively small, writing $r_j = \bar{r}.(1 + e_j)$, with $\Sigma e_j = 0$, and ignoring cubic powers of the e_j in the resulting power series expansion. The variability of the sample sizes is described to a first approximation by

$$\text{CV}_r^2 = \Sigma e_j^2/n_J. \tag{4.9}$$

Write

$$\rho_\epsilon^2 = (\tau_\epsilon/\bar{r}.)/(\tau_\xi + \tau_\epsilon/\bar{r}.), \quad \rho_\xi^2 = 1 - \rho_\epsilon^2, \tag{4.10}$$

describing approximately the proportions of the variance of the mean contributed by the two sources of variability. Then to the approximation indicated we have that

$$\text{var}(\bar{Y}_W) \;=\; (\tau_\xi/n_J + \tau_\epsilon/(n_J\bar{r}.))(1 + \rho_\xi^2 \text{CV}_r^2), \tag{4.11}$$

$$\text{var}(\bar{Y}_U) \;=\; (\tau_\xi/n_J + \tau_\epsilon/(n_J\bar{r}.))(1 + \rho_\epsilon^2 \text{CV}_r^2), \tag{4.12}$$

$$\text{var}(\hat{\mu}) \;=\; (\tau_\xi/n_J + \tau_\epsilon/(n_J\bar{r}.))(1 + \rho_\epsilon^2 \rho_\xi^2 \text{CV}_r^2). \tag{4.13}$$

This shows that, provided the relative variation in sample size is not extreme, the loss of efficiency from using one of the simpler estimates is not great so long as the correct choice between the two simpler estimates is made. This choice is to be based on which is the predominant source of variation in the sample means. In a sense the worst case is where $\rho_\epsilon^2 = 1/2$. Then the asymptotic relative efficiency of the simpler estimates is $(1 + \text{CV}_r^2/4)/(1 + \text{CV}_r^2/2)$.

A reasonable procedure is to estimate ρ_ϵ^2 by any convenient method and to estimate μ by $\bar{Y}_U$ or by $\bar{Y}_W$ according to whether ρ_ϵ^2 is less than or greater than one-half.

Loss of efficiency is conventionally and naturally expressed via the relation between estimates and the underlying parameter value. In immediate practice, however, it is revealed by how much a notionally inefficient estimate is changed when replaced by a corresponding efficient estimate. In general if $\hat{\theta}$ is an efficient estimate and T another estimate of a parameter θ then to the approximations involved in first-order asymptotic theory we can write $T = \hat{\theta} + Z$, where Z is independent of $\hat{\theta}$ and all random variables are normally distributed. It follows that

$$\text{st.dev.}(Z) = \text{st.dev.}(T)\sqrt{(1 - \mathcal{E})}, \tag{4.14}$$

where $\mathcal{E} = \mathrm{var}(\hat{\theta})/\mathrm{var}(T)$, the ratio of the asymptotic variances, is the asymptotic relative efficiency of T.

Thus if $\mathcal{E} = 0.9$ or 0.95 the factor takes values 0.32 or 0.22 so that, in the latter case, the change in the estimate in going from a 95% efficient estimate, T, to a fully efficient one would usually be less than one-fifth of a standard error of T and is very unlikely to be as much as one half the standard error.

In this discussion the focus of interest is the mean μ and the components of variance enter incidentally. There is a parallel but rather more complicated discussion for estimation of the τ's.

Estimation of τ_ϵ is, subject to its assumed constancy, unaffected by the lack of balance and is based on the residual mean square within groups. For simple estimation of τ_ξ there are several possibilities, the two most obvious ones being based on, respectively,

$$\mathrm{MS_W} = \Sigma r_j (\bar{Y}_{j.} - \bar{Y}_\mathrm{W})^2/(n_J - 1), \tag{4.15}$$

$$\mathrm{MS_U} = \Sigma (\bar{Y}_{j.} - \bar{Y}_\mathrm{U})^2/(n_J - 1). \tag{4.16}$$

The former is the standard analysis of variance sum of squares between groups. Then

$$E(\mathrm{MS_W}) = (r_. - \Sigma r_j^2/r_.)\tau_\xi/(n_J - 1) + \tau_\epsilon, \tag{4.17}$$

$$E(\mathrm{MS_U}) = \tau_\xi + \tau_\epsilon(\Sigma 1/r_j)/n_J, \tag{4.18}$$

from which, in conjunction with MS_ϵ, estimates of τ_ξ are found directly. The same considerations apply to negative estimates as in the balanced case.

To a certain extent parallel arguments hold for unbalanced multi-way and nested structures, at least so long as none of the cells of a multi-way lay-out is empty. Note also that in the generalization in which components of covariance are involved, i.e., in which τ_ϵ and τ_ξ are covariance matrices, no new computing formulae are needed. Covariances can be estimated from parallel computations on two variables and their sum.

4.3 A more general formulation

We now show the relevance of the simple and apparently rather special discussion of Section 4.2 to a quite wide-ranging set of problems. Let $T_1, \ldots, T_{n_J}$ be estimates of a parameter θ obtained from independent sets of data, each with an internal estimate of error. Suppose that the estimates vary more than would be expected on the basis of internal error and that it is not feasible to explain the extra variability as systematic, for example by regression on whole-set explanatory variables. Then we may represent the additional variability as random and this leads, via the usual approximations of first-order asymptotic theory, to

$$T_j = \theta + \xi_j + \epsilon_j, \tag{4.19}$$

where the errors ϵ_j, ξ_j are approximately normally distributed and are independent. Here ϵ_j represents an internal error of estimation of zero mean and variance $\tau_\epsilon r_j^{-1}$ so that the parameter value for the jth group of data is $\theta + \xi_j$. Thus ξ_j assumed of zero mean and variance τ_ξ represents between-group variation in the parameter.

In the following discussion we shall suppose that τ_ϵ is estimated from a large number of degrees of freedom or equivalently known *a priori*. Thus we may without loss of generality set $\tau_\epsilon = 1$, or any other convenient constant. Thus the r_j are no longer sample sizes and in general not integers.

For example, suppose that for the jth group of data a model is fitted. say by maximum likelihood, leading to an estimate T_j of a scalar parameter θ for that set of data. Then r_j^{-1} can be obtained from the inverse of the observed information matrix. Note that it is not necessary that the same model is fitted to each group of data. only that the parameter θ have the same interpretation throughout. For example, in combining the results of a number of case-control studies. the parameter θ could be the log odds ratio for treatment versus control estimated after adjustment by logistic regression for imbalance with respect to covariates that might be different in the different studies.

The procedure suggested here is first to analyse each group of data separately recording estimated parameters and their internal standard errors. Assuming for the moment a scalar parameter of interest, this leads to a set of estimates, T_j with estimated standard errors and the equivalent r_j. It will then require only simple analysis to decide whether a component of variance τ_ξ is necessary, whether there are apparently outlying groups and which of the two relatively simple combined estimates of θ is likely to have high efficiency. The formal fitting of a model with random regression terms may be quite complicated and, possibly, less insightful.

If the parameter θ is a vector the above arguments apply with minor change. Thus r_j is the internal information matrix for θ from the jth group and τ_ξ is a covariance matrix between groups.

A typical relatively informal application of these ideas might proceed as follows. To each group of data a multiple regression model is fitted: special cases are ordinary normal-theory linear regression, linear logistic regression, etc. For a single combined analysis it may be helpful to first examine the internal and between-group variances of the estimated intercepts and regression coefficients to find those aspects for which a random between-group component is likely to be important. Of course if appreciable variation between groups is found in an important feature, a substantive explanation is preferable to description as a random variable. If a representation with several kinds of random variation is chosen, approximately efficient estimation of averaged parameters and of variances follows the principles outlined above. This may be followed by nominally efficient estimation via one of the packages for multi-level modelling; the present approach avoids the use of possibly over-elaborate models and shows the circumstances under which the conclusions from elaborate models are, or are not, virtually the same as from simpler approaches.

If in a regression problem a randomly varying intercept and a randomly varying regression coefficient are both necessary it will nearly always be essential to allow these random terms to be correlated. For instance, in the simplest case of normal-theory linear regression of, say, Y on a single explanatory variable z with groups indexed by j, the model

$$Y_{js} = (\theta_0 + \xi_{0j}) + (\theta_1 + \xi_{1j})z_{js} + \epsilon_{js}, \tag{4.20}$$

with ξ_{0j} and ξ_{1j} assumed to be independent random variables represents only a very restricted form of random variation in the individual within group regression lines. The latter model is, in any case, not invariant under changes of the origin of z.

4.4 A special case

We now consider in a little more detail a special case in which within the jth group, assumed not too small, there is linear logistic regression of a binary response Y_{js} on a single explanatory variable z_{js}. Write $L(x) = e^x/(1+e^x)$, the unit logistic function. Even in this simple case there are quite a number of models that might be appropriate should a single logistic regression, the same for all groups, fail. The following list is not exhaustive: we exclude, for example, models not on a logistic scale and models of nonlinear logistic regression. If there is more than one explanatory variable the discussion can be extended.

Single logistic regression. Here

$$P(Y_{js} = 1) = L(\beta_0 + \beta_1 z_{js}), \tag{4.21}$$

a single logistic regression with the same intercept and slope for all groups. This and similar 'fixed-effect' models can be fitted by maximum likelihood. Sometimes, for grouped data, the asymptotically equivalent technique of empirically weighted least-squares analysis of the empirical logistic transform, an old-fashioned technique, still valuable from time to time, is a convenient alternative.

Parallel logistic regressions. This differs from the previous model only by replacing β_0 by β_{0j}. With a modest number of groups it may often be best to regard the intercepts as separate unknown parameters and to continue use of maximum likelihood, or for analysis of the slopes in principle the conditional likelihoods given the total numbers of successes per group, $Y_{j\cdot}$.

Fixed intercept. The formal equivalent of the previous case with fixed β_0 and variable slope will be useful only if the origin $z = 0$ has a very particular subject-matter interpretation. This is because all the logistic curves coincide at this single point.

Arbitrary logistic regressions. If we combine the first two cases we obtain separate regressions for each group of data.

Whole-group explanatory variables. If there is substantial variation in one or both logistic regression coefficients explanation in terms of a whole-group explanatory variable is desirable, writing, for example,

$$\beta_{1j} = \beta_1 + \gamma_1 w_j. \tag{4.22}$$

We now turn to models with a second layer of random variation.

Random intercepts. This is obtained from the model of parallel logistic regressions by supposing the intercepts to be random variables. i.e.. by writing

$$\beta_{0j} = \beta_0 + \xi_j. \tag{4.23}$$

We have, essentially by definition of β_0. that $E(\xi_j) = 0$. We write as before τ_ξ for its variance. Any assumption of normality for ξ_j will be difficult to check except from very extensive data. One advantage of this model over the earlier model is that τ_ξ provides a concise summary of the instability in the intercept which may be useful in comparing different sets of similar data. Another is that if interest focuses on a particular group empirical Bayes shrinkage of estimates becomes possible.

A model derived in the formally analogous way by confining random variation to the slope β_{1j} is unlikely to be of interest; it implies that at a known value of z, taken to be the origin. there is no random variation between groups in the underlying probability.

Random regression model. Here we take the model in which the logistic regression parameters are written in the form

$$\beta_{0j} = \beta_0 + \xi_{0j}. \quad \beta_{1j} = \beta_1 + \xi_{1j}. \tag{4.24}$$

where now τ_ξ denotes the covariance matrix of (ξ_{0j}, ξ_{1j}).

The models (4.23) and (4.24) can be fitted using available software routines for logistic regression with random effects, the latter under a largely untestable normality assumption. From the present viewpoint they can be analysed more simply as follows: we take (4.24) as an example.

The steps are as follows:

(i) fit separate logistic regressions to each group of data:

(ii) tabulate the estimates $(\hat{\beta}_{0j}. \hat{\beta}_{1j})$ and their estimated covariance matrices;

(iii) assess whether the variation between groups in the estimates is predominantly from internal variability:

(iv) estimate (β_0, β_1) by the appropriate simple combination, weighted or unweighted, of the separate group estimates.

The advantage of the simple procedure is perhaps greater in more complex problems, for example those of multiple logistic regression.

As is well known, care is needed in interpreting regression parameters in models that are not linear in the expected value of the response random variable. Logistic regression again illustrates a general argument. In (4.21) the parameter β_1 has the following interpretation. For an arbitrary individual the logit of the probability of success is increased on the average by β_1 for a unit increase in z. Under the random intercept model this is, however, not the same as the difference of the logit probabilities of success for two randomly chosen individuals with z differing by one, the so-called marginal interpretation. This is because marginally

$$P(Y_{js} = 1) = E\{L(\beta_0 + \xi_j + \beta_1 z_{js})\}, \tag{4.25}$$

where the expectation is over the distribution of ξ_j.

One way of approximating (4.25) (Cox, 1966) is by approximating $L(.)$ by a normal integral, evaluating the expectation exactly, in fact as another normal integral, and then approximating back as the logistic function. This gives

$$P(Y_{js} = 1) = L\left(\frac{\beta_0 + \beta_1 z_{js}}{\sqrt{(1 + c^2 \tau_\xi)}}\right), \tag{4.26}$$

where $c \simeq 1.65$ is the scaling constant involved in the correspondence between the unit logistic and the standardized integrated normal functions.

An alternative argument, applicable much more generally, is via Taylor expansion to give

$$P(Y_{js} = 1) \simeq L(\beta_0 + \beta_1 z_{js}) + L''(\beta_0 + \beta_1 z_{js})\tau_\xi/2 \tag{4.27}$$

and because of the special properties of the function $L(.)$ this reduces to

$$L(\beta_0 + \beta_1 z_{js})[1 + \{1 - 2L(\beta_0 + \beta_1 z_{js})\}\tau_\xi/2]. \tag{4.28}$$

Example 4.1. Australian intensive care outcomes. The Australian and New Zealand Intensive Care Society (ANZICS) maintains a voluntary database of adult patient intensive care outcomes from Intensive Care Units (ICUs) across Australia. In this example, we analyse data on 71,578 patients from 109 ICUs over a 52-month period from 1993 to 1997 for which relatively complete patient information and hospital data are available. The data are highly unbalanced with the number of patients per ICU site, r_j, ranging from 26 to 3318. A detailed analysis of these data is being undertaken in collaboration with J. Moran (The Queen Elizabeth Hospital, Adelaide) and D. Firth (Nuffield College, Oxford).

The outcome of interest is whether or not patients were alive when they left the hospital, and in the present analysis we study the predictive value of individual patient characteristics, surgical and other hospital-related information. There is some degree of wooliness in the mortality indicator in that, for example, patients who were transferred to another hospital may have been mortally ill and died during transfer, or at the second site. Such information is not recorded in the database, which point of course is a not atypical limitation of observational databases of this type.

Individual patient data include demographic variables such as age and sex, hospital length of stay, several diagnostic and prognostic patient scores. clinical diagnostic codes at admittance to intensive care, ICU site and geographic designation: one of rural, metropolitan, tertiary or private. There is geographical variation amongst the hospitals in that patient, nursing and medical staff can be of quite different case mix.

The most important and widely used prognostic score in clinical intensive care is the Acute Physiology and Chronic Health Evaluation II (APACHE II) score. This is a clinically derived point score for each patient based on:

(i) 12 routine physiological measurements (sodium, potassium. creatinine, white cell count, haemoglobin. Glasgow coma score, temperature. mean arterial pressure, heart rate. respiratory rate, arterial pH and oxygenation index) scored from 4 to -4 as derived from clinical impression:

(ii) age in years categorized as $\leq 44, 45-54, 55-64, 65-74, \geq 75$ scored 0,2,3,5 and 6 respectively; and

(iii) chronic health points: if the patient has a history of severe organ dysfunction or is immuno-compromised, points are assigned as follows: non-operative or emergency post-operative patients scored 5, and elective post-operative patients, 2.

The APACHE II score is equal to (i) + (ii) + (iii) with a range from 0 to 71, and a clinical upper limit of approximately 55. In Australia. the average score for tertiary-referral ICUs is approximately 17.

The APACHE II *algorithm* is the most widely used algorithm for routinely computing risk of death or probability of death in the ICU literature. It is based on the APACHE II score described above, post-operative emergency surgery status, plus 50 clinical diagnostic categories. Currently. there is some debate about the statistical usefulness or otherwise of the model for the logit of APACHE II risk of death, from which standardized mortality ratios (SMRs) are derived to form the basis of 'league tables' for ICU performance. Controversy surrounds the issue of the validity of the SMR to the extent that statistically significant departures from a value of one represents a true indication of excess or decreased mortality. For instance. it could be argued that the algorithm is flawed because there is no adjustment for case mix, and that the SMR only has meaning after examining all aspects of an ICU's performance. We will not pursue this debate here.

but focus on modelling the predictive performance of the APACHE II score and other explanatory variables directly via binary logistic regression. See the Bibliographic notes for key references and further information about APACHE.

In the first instance we evaluate the performance of APACHE II and hospital length of stay in predicting hospital mortality. Since age is already incorporated into APACHE II, it is not included as a separate explanatory variable in our analysis.

Gender has been shown to have no association with mortality in this context, and our initial analysis also established comparable results for males and females; thus gender is not considered further here either. For our illustrative purposes, we do not consider the emergency surgery indicator or the diagnostic categories utilized in the APACHE II algorithm, but acknowledge that these may be important predictors of hospital mortality.

We note that duration of stay may be influenced by the current assessment of possible imminent death and that this biases interpretation; duration is also highly right-skewed with a small number of very long times. The mean, standard deviation and median duration of stay in hospital are 21.75, 310.67 and 9.90 days, indicating that some patients spend a very long time in hospital.

To handle this feature of the data, we adapt a suggestion by Cox and Wermuth (1996) to split the length of stay variable into two explanatory variables. The first is an indicator variable for whether the patient stayed in hospital less than one week (Dweek=0) or longer (Dweek=1), and the second variable (diffdays) is the number of days spent in hospital more or less than the baseline of one week, and therefore represents the extent of departure from baseline. Although there are problems inherent in the use of cut-points to parameterise continuous explanatory variables, the representation used here avoids an analysis which may be too dependent on the precise values of the very short times, which is also an argument against the use of transformations such as reciprocals or logs. We discuss briefly the controversy over the application of cut-points to summarise continuous variables in the bibliographic notes.

We centre the explanatory variables about their means in order to aid quantitative interpretation of the results and, in particular, to avoid misinterpreting the behaviour of the intercept relative to the other explanatory variables.

Single logistic regression. To begin, we fitted a global binary logistic regression model, for the moment ignoring site (i.e., hospital), and including the APACHE II score and hospital length of stay variables described above, plus possible interaction terms. The total observed proportion of hospital deaths is 0.19. The results from the best fitting single regression model are set out in Table 4.1, where, as expected with such a large number

Table 4.1 *A single logistic regression model fitted to the ANZICS data. Each explanatory variable is centred about its mean.*

Variable	$\hat{\beta}$	st. err.
Intercept	−3.86	8.30×10^{-2}
APACHE II	0.18	1.71×10^{-3}
Dweek	4.52	2.21×10^{-1}
diffdays	−0.11	4.40×10^{-3}
APACHE II:Dweek	−0.14	3.85×10^{-3}
Dweek:diffdays	0.28	1.18×10^{-2}

of patients, the effects are formally highly statistically significant. We do not attempt a quantitative interpretation of these preliminary results. but include them at this stage to motivate the ensuing model in which these explanatory variables are fitted separately to the data from each hospital.

The probability of death in hospital increases with increasing APACHE II score: the overall mean score is 13.9 with standard deviation 8.8. In a naive interpretation of the main effects. hospital stays of longer than one week (Dweek=1) are associated with an increased probability of death. with the change in probability depending on the extent of the departure from baseline. In particular, very long hospital stays (and there are some extreme values) are associated with decreased probability of death. It is worth noting at this point that if diffdays is omitted from the model. the Dweek indicator has a negative (estimated) coefficient and some sensitivity in the interpretation of the hospital length of stay variables may be lost. The interaction between the two ·time· explanatory variables represents two-phase regression. The interaction of Dweek with APACHE II suggests that the predictive value of APACHE II is attenuated in patients who stay in hospital for periods longer than one week.

We investigated possible nonlinear effects in APACHE II and diffdays using generalized additive models in the global model described in Table 4.1. Figure 4.1 shows the smoothed terms respectively for the centred variables APACHE II (here denoted ap2mm) and the difference in days from the baseline of one week in hospital (denoted diffdaysm). and demonstrates reasonable linearity for diffdays ($P = 0.06$). There is some nonlinearity in the APACHE II score ($P = 0$), with a slight bulge between scores of about 12 and 30. One possibility would be to fit a quadratic term in APACHE II. but clearly this would not capture the linear effects at the lower and higher scores. Since the nonlinearity is relatively mild in real terms, we retain the APACHE II score as a linear predictor.

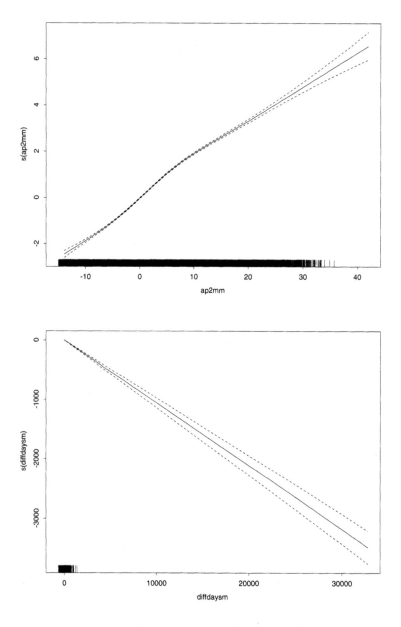

Figure 4.1 *Checking nonlinearity in the continuous explanatory variables in the single logistic regression model in Table 4.1: generalized additive models fitted to the centred variables APACHE II score (ap2mm, top picture), and diffdays (diff-daysm, bottom picture), the latter representing time spent in hospital departing from the baseline of one week. The dashed lines are at plus and minus two point-wise standard deviations. The rug along each horizontal axis marks the observed values for the explanatory variable.*

The rugs on the horizontal axes in the plots mark the observed values for each explanatory variable. For APACHE II, there is a small number of high, extreme values, but otherwise the observations are dense at all scores up to 45 or so. By contrast, most of the diffday values are very small and dense around zero with a few extremely large values.

Given the large number of hospital sites under study, we did not fit parallel logistic regressions separately to each hospital. Of course it is entirely plausible that in a different context or application, interest may focus on individual hospitals which would then be represented by separate unknown parameters.

Random regression model. Using the model in Table 4.1 as a basis for further investigation, we explore the potential of more complex models by extending the formulation for random intercepts and slopes (4.24) to multiple logistic regression. Our aim in the first instance is to determine which of the two extreme forms of regression parameter estimates analogous to (4.5) and (4.6) is appropriate for the ANZICS data, and thus to establish whether the internal or between-hospital variability is dominant in the individual explanatory variables.

Separate logistic regressions were fitted to each of the 109 hospitals. and the estimated coefficients together with their estimated covariance matrices were output to a file.

Let $\bar{T}$ be the sample mean for a given regression estimate. for example. the intercept. Then the unweighted analysis is based on

$$E\left\{\frac{1}{n_J-1}\sum_{j=1}^{n_J}(T_j-\bar{T})^2\right\} = \tau_\epsilon + \frac{1}{n_J}\sum_{j=1}^{n_J}s_j^2, \qquad (4.29)$$

where s_j^2 is the estimated variance of the regression coefficient obtained from the observed variance-covariance matrix.

The weighted analysis is based on the overall mean

$$\tilde{T} = \frac{\sum_{j=1}^{n_J}T_j/s_j^2}{\sum_{j=1}^{n_J}1/s_j^2}, \qquad (4.30)$$

with, approximately,

$$E\left\{\sum_{j=1}^{n_J}\left(T_j-\tilde{T}\right)^2/s_j^2\right\} = \tau_\epsilon\left(\sum_{j=1}^{n_J}1/s_j^2 - \frac{\sum_{j=1}^{n_J}1/s_j^4}{\sum_{j=1}^{n_J}1/s_j^2}\right) + n_J - 1$$

$$\qquad (4.31)$$

We do not need to make assumptions about the form of the internal error of estimation, other than that we treat the standard errors of the estimates as known constants. In each case, the expectation is equated to its observed value and the resulting equation solved for τ_ϵ.

Table 4.2 *Unweighted analysis of separate logistic regressions fitted to each hospital from the ANZICS database; $n_J = 109$ hospitals: estimated means and standard errors.*

Variable	$\bar{\hat{\beta}}$	st. err.
Intercept	−3.60	2.05×10^{-1}
APACHE II	0.19	6.02×10^{-3}
Dweek	3.91	4.89×10^{-1}
diffdays	−0.09	1.01×10^{-2}
APACHE II:Dweek	−0.13	1.03×10^{-2}
Dweek:diffdays	0.25	2.77×10^{-2}

A problem that arose in fitting the separate logistic regressions by maximum likelihood is the phenomenon known as 'monotone likelihood' or 'complete separation effects' where the responses death (=1) are perfectly separated from the non-deaths (=0) by either a single explanatory variable or by a combination of variables. Then the fitted values are arbitrarily close to zero and one and the estimation procedure does not converge. In one sense, this is a desirable outcome as it indicates that the model is a good one, although it could also be viewed as a form of model over-fitting and unlikely to represent real total separation in the population. It is important to include such hospitals in the analysis in order to avoid introducing bias into the results which could result from their omission. To cope with the problem, we employed a fitting procedure proposed by Firth (1993) for reducing the bias in maximum likelihood estimates in generalized linear models (see the Bibliographic notes and Computational/software notes for further details).

Tables 4.2 and 4.3 set out the results of the empirical unweighted and weighted analyses, respectively. For each analysis, we note the similarity of the resulting parameter estimates and estimated standard errors of the regression coefficients to the single logistic regression model summarised in Table 4.1.

We may draw several tentative but important conclusions from the results presented in Tables 4.2 and 4.3, together with the unweighted variance component analysis (not shown). The estimated unweighted components of variance were found to be typically small and negative, with the exception of that for APACHE II score, which was small and positive (namely, $\hat{\tau}_{\xi}^{U} = 0.00151$). The only large negative estimate was that for Dweek, which had an estimated variance component slightly smaller than −10. Clearly, the overall dominant sources of variability are internal to hospitals rather than between hospitals, and thus the weighted analysis is the more appropriate here. One way to think about this is that if the internal variability

Table 4.3 *Weighted analysis of the separate logistic regressions fitted to each hospital from the ANZICS database; $n_J = 109$ hospitals. E(SS) is the left-hand side of (4.28). Var: variable; Int: intercept; A: APACHE II; D: Dweek; diff: diffdays; s.e.: standard error.*

Var	$\tilde{\hat{\beta}}$	s.e.	$E(\hat{S}S)$	$\hat{\tau}_\xi^W$	$\sqrt{\hat{\tau}_\xi^W}$
Int	-3.65	5.50×10^{-2}	221.07	8.97×10^{-1}	9.46×10^{-1}
A	0.18	2.44×10^{-5}	343.39	8.87×10^{-4}	2.98×10^{-2}
D	4.11	3.91×10^{-1}	156.96	2.76×10^{-0}	1.66×10^{-0}
diff	-0.10	1.58×10^{-4}	166.97	1.35×10^{-3}	3.67×10^{-2}
A:D	-0.12	1.24×10^{-4}	156.95	9.22×10^{-4}	3.04×10^{-2}
D:diff	0.27	1.13×10^{-3}	171.94	1.05×10^{-2}	1.02×10^{-1}

for an effect is small, then it would make sense to take the group means to represent the true mean of the population.

The unweighted means of the regression coefficients in Table 4.2 are nevertheless of interest. Taken together with their estimated standard errors. the results suggest that Dweek, diffdays and their associated interaction terms are perhaps less important as explanatory variables than APACHE II and the intercept representing hospital site.

From the weighted analysis summarised in Table 4.3, it is apparent that both the intercept and Dweek indicator have substantial components of variance between hospitals and are potential candidates for fitting as random effects. The remaining four components of standard deviation are all rather small, and APACHE II, together with its interaction with Dweek. can reasonably be assumed to be fixed effects. This is also true for diffdays. for which the estimated variance component is negligible.

It is plausible *a priori* that at least some of the variation observed in the intercept and Dweek is due to variation between hospitals and their associated patient case mix, medical facilities and medical staff. We investigate this further below, following a brief examination of the components of covariance.

The covariance components, or more particularly, the correlation components, may help shed light on the behaviour of the substantial effects observed in the data. For the unweighted analysis, the components of covariance can be obtained simply from the estimated variances of $\tilde{\hat{\beta}}_j$ and $\tilde{\hat{\beta}}_l$, $j \neq l$, together with the variance of their sum, say. However. since the weighted analysis is appropriate for these data, we consider only the weighted components of covariance for the $\tilde{\hat{\beta}}$.

The relatively small estimated components of variance for APACHE II. diffdays and the interaction terms induce some instability in the empirically

Table 4.4 *Weighted means and components of standard deviation from the separate logistic regressions fitted to the hospitals in the ANZICS database stratified by four hospital geographical designations.*

		Intercept			Dweek		
	n	$\tilde{\beta}$	s.e.	$\sqrt{\tilde{\tau}_\xi}$	$\tilde{\beta}$	s.e.	$\sqrt{\tilde{\tau}_\xi}$
Regional	25	−3.48	0.12	1.09	3.89	0.86	1.77
Metropolitan	21	−4.01	0.09	0.72	5.12	0.64	1.47
Tertiary	33	−3.66	0.11	1.09	3.83	0.75	1.62
Private	17	−3.24	0.14	1.17	3.06	1.02	2.66

estimated correlation components, some of which have estimated values greater than one. (Of course, theoretical correlations cannot be greater than unity and such errant estimates serve as a warning that, in practice, the variance components need to be rather precisely determined.) Dweek is highly negatively correlated with the intercept ($r = -0.87$), and again it is plausible that a substantial fraction of the variation observed in both the intercept and in Dweek is due to the case mix in the different hospitals. We now explore this feature of the data in more detail.

Table 4.4 gives the weighted means for the explanatory variables intercept and Dweek by hospital geographical designation. Some systematic and random features are immediately apparent.

For a given APACHE II score, the different hospital levels have similar intercepts, with metropolitan hospitals having a slightly lower average intercept compared with the overall weighted mean. At the same time, Dweek indicates that staying in a metropolitan hospital for more than a week is associated with an increased risk of death. Private hospitals, which are the smallest group, have a smaller average Dweek associated with a reduced probability of death, but the hospital types are otherwise similar. As before, it is highly likely that the case-mix differences which (at least approximately) distinguish the hospital levels underlie these features of the data.

The results given in Table 4.4 also suggest the existence of substantial components of variance for both the intercept and Dweek indicator for a given APACHE II score, with similar effects within each hospital level. Overall, the intercept components of standard deviation are similar and close to one, although metropolitan hospitals are relatively less variable than the other hospital types. The components of standard deviation for Dweek are typically larger for the four hospital levels, and very similar for regional, metropolitan and tertiary-referral hospitals. Private hospitals on the other hand appear to be relatively more variable with regard to

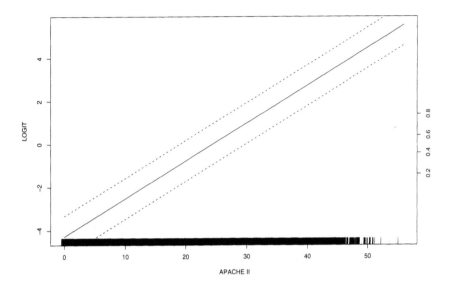

Figure 4.2 *Fitted logits and fitted probabilities of mortality from the weighted analysis as a function of APACHE II, over the range of scores observed in the ANZICS data (from 0 to the largest score of 56). Solid line: all explanatory variables and the intercept fixed at their means; dashed lines are plus or minus one standard error in the intercept. The rug of values is the distribution of the individual APACHE II scores; note that the predicted curves extend only over the range of the data with the fitted probabilities ranging from 0.012 to 0.997.*

the effect of hospital length of stay on the risk of death. The estimated component of correlation between the intercept and Dweek remains large and negative for each hospital type for a given APACHE II score, implying that hospitals with increased overall risk of mortality also have an associated reduced risk of death for patients who stay longer than one week.

From these results, we may conclude that incorporating hospital geographical region as a (fixed) whole-group explanatory variable is unnecessary and unlikely to explain the observed heterogeneity between hospital types.

The primary aims of this analysis of the ANZICS data have been to illustrate the methods propounded in the earlier sections of the chapter, and to aid in understanding the behaviour of important features of the data. We have also sought to obtain an interpretable, realistic model for the intensive care data, and one which will aid in the choice and interpretation of a global mixed effects logistic regression model. A further major aim has been to provide useful quantitative predictions from the model. With this in view, Figure 4.2 plots the logit probabilities of death in hospital as a function of increasing APACHE II score, from which it is evident that APACHE II is highly predictive of outcome. The probability scale on the vertical right-hand side gives the predicted probability of death for a given APACHE II score at admission to intensive care. The dashed lines at plus or minus one standard error in the intercept give the predicted probabilities for hospitals lying one standard deviation on either side of the mean.

In summary, the APACHE II score provides a remarkably stable predictor of mortality, but if a particular hospital wishes to translate this into an estimated probability, inclusion of the 'intercept' specific to that hospital is important. This is even more so the case for the Dweek indicator.

A single mixed effects logistic regression model: comment. On the basis of the analysis so far, one could now proceed to fitting more sophisticated models with some confidence. An obvious candidate would be a global logistic regression model incorporating the intercept and Dweek as correlated random effects. The interaction term APACHE:Dweek would then be random as well, even though we have already established that the interaction component of variance is very probably very small, and that a reasonable working approximation is that the interaction term is determined by the APACHE score and Dweek in the same way for all hospitals. Similar considerations apply to the Dweek:diffdays interaction term. This leads to a global mixed effects logistic regression model with covariance structure represented by a 4×4 covariance matrix incorporating the associated components of covariance.

The extremely variable number of patients per hospital merits further study as well, one possibility being to include $\log r_j$ as an explanatory variable in the model.

Although it would in principle be good to compare our empirical approach with a global random effects model, the approaches we tried ran into computational problems because of the large size of the dataset and the relative complexity of the model, including as it does several correlated random effects. We discuss this further in the Computational/software notes at the end of the chapter.

We may conclude that without the benefit of the simple weighted analysis outlined above, analysing these data and, in particular, determining which explanatory variables should be interpreted as random effects would have been a much more challenging task.

4.5 Synthesis of studies

A primary focus of much discussion of the statistical aspects of the design and analysis of studies is on investigations for which, so far as is achievable, a self-contained analysis of each study is feasible. Yet in many fields the synthesis of information from several or many investigations is crucial. Sometimes this can be achieved informally by careful review of the relevant literature. In others, more explicit statistical analysis of the combined data from a number of studies is helpful. From the point of view of the present book this introduces another feature, variation between studies and interactions of such variation with the treatment effects under investigation and each may involve representation by a component of variance.

The issues involved probably have a long history but seem first to have been studied systematically in the context of agricultural field trials under the broad heading of the analysis of series of experiments. More recently the emphasis has shifted to other fields including medical statistics. The term *overview* or *meta-analysis* is often used in this context; it should be emphasized, however, that once the data for analysis have been chosen no essentially new principle of analysis arises.

There are a number of rather different reasons why such combined analyses may be important. One is where the conclusions are to be applied in a range of circumstances.

Illustration. The conclusions from agricultural field trials may be applied in ranges of soil and meteorological conditions. Thus a very precise study on one farm and in one year may be a poor basis for a general recommendation. Replication in a number of farms and in a number of years, the latter to examine variation with meteorology, is therefore highly desirable. There are broadly similar considerations in connection with clinical trials: studies in different centres and different hospitals allow for different patient mixes and other features which may influence a treatment assessment.

Another reason is particularly relevant when relatively small effects are under study. Synthesis of different studies may be necessary to achieve the level of precision desirable. Also, even if the individual studies are large, the sensitivity to bias is rather less when independent studies are combined. It is, for example, not impossible that all of a number of independently designed case-control studies are subject to a common bias from an unobserved confounder, but this is less likely than in a single very large study of common design.

The reliance on the synthesis of information from several studies does not remove the need for careful design of individual studies.

Illustrations. In an industrial context a series of trials of a new or modified process in a number of factories might show promising results in most but a very poor result in one might well lead to the modification being abandoned. In tests of a possible carcinogen by a number of routes, one positive result might lead at the very least to substantial extra work, of little value if the initial conclusion was defective. In these and similar situations one anomalous result arising from poor design or the play of chance may have substantial impact.

A key feature of any major synthesis is the choice of studies for inclusion. Criteria include the quality of the study, in so far as it can be judged, the subject-matter comparability of the contrasts under study and completeness of the set of studies to be included. The last aspect is important because, for example, any tendency not to publish studies which have shown no statistically significant effect would bias the overall conclusions based on published studies. We shall assume in the following discussion that all studies lead to estimated treatment effects on a common scale and may therefore be expected to have similar numerical values.

In the application to agricultural field trials and to other systems which are really planned as a single entity, it may be that the same design is used at all sites and times, even though the studies are in a sense statistically independent. As with other fully balanced sets of data the formal analysis is simplified by the balance induced, although in principle the same ideas apply much more generally.

Apart from these considerations for the inclusion of data, the principles of analysis are those set out in previous chapters.

Suppose then that we have the comparison of n_T treatments at n_P studies, for example in different places, and that, possibly after eliminating effects such as blocks in a randomized block design, the residual error is effectively constant and that the same design is used in all studies. Then the resulting balanced data have an analysis of variance, the relevant parts of which are as follows:

$$
\begin{array}{ll}
\text{Between treatments} & n_T - 1 \\
\text{Between studies} & n_P - 1 \\
\text{Treatments} \times \text{studies} & (n_T - 1)(n_P - 1) \\
\text{Pooled residual} & n_P n_{\text{res}}
\end{array}
$$

where n_{res} is the number of residual degrees of freedom per study.

There are now a number of possibilities. If the interaction shows clear evidence of variation in treatment effect between studies, then we should

- find a rational explanation of the nonconstancy, for example by a transformation of the scale of measurement or via whole-study explanatory variables

- check whether the nonconstancy is largely confined to certain degrees of freedom within the space of treatment contrasts, for example contrasts confined to a particular individual treatment or to perhaps to the slope of the dependence on a factor with quantitative levels

- check at least informally that the variation in treatment effect is large enough to be of subject-matter importance, bearing in mind that in many contexts complete constancy of treatment effects is *a priori* unlikely

If no direct explanation of the variation in treatment effect along the above lines is available, it may then be necessary to regard that variation as an additional source of random variation with its associated component of variance. That is, we represent an observation on treatment t in study p in the form

$$Y_{tps} = \mu + \theta_t + \gamma_p + \xi_{pt} + \epsilon_{tps}, \tag{4.32}$$

where θ_t characterizes the tth treatment averaged over the population. γ_p is a study effect and the zero-mean random variables ξ_{pt}, ϵ_{tps} of variances τ_ξ, τ_ϵ represent, respectively, treatment $\times$ study interaction and internal variability.

Now if this model is taken as the basis for interpretation, then the discussion earlier in the book applies and the variance of a treatment contrast is estimated via the mean square for interaction. Provided there are enough studies to generate sufficient degrees of freedom for that mean square to be estimated with reasonable precision there is no special difficulty.

But what does such a model mean? If the studies are in some sense like a random sample from a population of studies the interpretation is clear. The parameters θ_t refer to an average over that population. In practice. even in the agricultural context mentioned above. it is unlikely that the studies. i.e., experimental farms, are effectively a random sample of prospective user farms. In a clinical trial setting it is unlikely that centres taking part in such a set of studies are like a random sample from a target population. A weaker interpretation is that the random sampling refers only to the interaction terms in the following sense. The treatment contribution varies from study to study. If we are unable to explain that variation in systematic terms we may provisionally assume that the variation is generated by processes producing effectively random variation. Our target parameter averages over that process. Note that for the present purpose the randomness or otherwise of the parameters γ_p is irrelevant. This latter argument is essentially used in standard discussions of error estimation in experimental design when. for example, interaction of treatments and blocks in a randomized block design is used to estimate error without any specific assumption that the blocks are samples from a meaningful population.

If we defined the target parameter as an average over the studies actually available the error contribution would come only from the random variables ϵ estimated by the pooled residual mean square within studies. Usually this will be less than, and possibly much less than, the interaction mean square.

If, as sometimes happens, the study variable refers to time. for example to an experiment replicated in a number of years, there is the further complication that the additional variability may be serially correlated in time. or show time trends or other systematic features.

If the studies are not of the same design, and in particular lead to individual estimates of very different precisions. there are the issues connected with unbalanced data discussed above. In particular, if the studies are of very different sizes, so that the internal precisions of the different estimates are very different, introducing the random interaction term moves the point estimate away from one in which each study is weighted inversely as the variance of the corresponding estimate, i.e., roughly weighting by sample size, towards the estimate in which each study is given equal weight regardless of sample size. Further there will be an increase, often substantial, in the variance of the overall estimate.

The following additional objection has been raised to uncritical use of the model (4.32). Suppose that a number of studies all show clear positive effects but that the magnitudes of the effects are quite different in the different studies. There will thus be a substantial interaction component of variance. This combined with a normal theory assumption may imply that there is an appreciable chance of a future study showing a negative effect. Yet it might be argued, partly on the basis of subject-matter knowledge, that although the magnitude of the effect to be encountered in a future study may be unpredictable, it is highly likely to be positive. That is, the treatment by study interaction is unlikely to be qualitative, i.e., to show effect reversals. In that case the analysis ignoring the additional component gives the correct idea of the strength of evidence in favour of a positive effect, although of course not of the predicted magnitude of effect. In principle the issue could be dealt with via appropriate assumptions of distributional form. Note that this argument would completely fail if, in a medical context, the interaction arose from different patient mixes with, say, a new treatment favourable for some patients and harmful for others.

Example 4.2. Bladder cancer and genotype. The data in Table 4.5 bring together evidence from 21 case-control studies examining the possible effect of NAT 2 (*N*-acetyl transferase 2) on bladder cancer. NAT 2 occurs in one of two forms, slow and fast. For a full discussion including the limitations of the data, see Green et al. (2000); for example, only five of the studies provided information on smoking status, despite that being a substantial risk factor for bladder cancer. The combination of data from different case-control studies generally encounters the difficulty that the control groups in the separate studies may not be comparable.

Evidence for an association between genotype and bladder cancer would be shown by a systematic departure from one in the odds ratio in the 2×2 contingency tables. We analyse this by the methods set out in Section 4.3.

For the jth table we denote by T_j the log odds ratio, i.e., the log of the product of the diagonal elements divided by the off-diagonal elements. Thus for the first table $T_1 = \log\{(233 \times 158)/(215 \times 141)\} = 0.1942$. Then T_j has an asymptotic variance v_j, say, the sum of the reciprocals of the individual frequencies; thus $v_1 = 1/233 + \ldots + 1/141 = 0.0224$. Table 4.5 summarises this first stage of analysis.

Table 4.5 *Summary of 21 case-control studies from Green et al. (2000); for each cell, case/control frequencies, estimated log odds, T_j, and variance v_j.*

Study, j	NAT2 slow	NAT2 fast	T_j	$10^2 \times v_j$
1	233/215	141/158	0.1942	2.24
2	154/135	100/107	0.1993	3.81
3	121/109	109/94	−0.0272	3.71
4	145/54	83/46	0.3976	5.92
5	127/26	62/33	0.9555	9.28
6	83/90	47/67	0.2736	5.94
7	80/79	35/39	0.1208	7.94
8	76/26	38/20	0.4308	12.79
9	74/54	37/41	0.4177	8.34
10	65/18	40/24	0.7732	13.77
11	67/149	35/143	0.6082	5.72
12	66/510	34/342	0.2637	4.94
13	59/54	39/56	0.4503	7.84
14	16/14	82/70	−0.0217	16.04
15	46/28	31/52	1.1014	10.89
16	20/13	51/78	0.8557	15.94
17	46/38	25/36	0.5557	11.58
18	47/10	20/12	1.0367	25.46
19	5/11	57/71	−0.5688	32.25
20	12/13	41/88	0.6837	19.60
21	3/13	48/190	−0.0905	43.64

First, application of the method of weighted least squares under a model of homogeneity leads to the estimated log odds of

$$\Sigma T_j/v_j/\Sigma 1/v_j \qquad (4.33)$$

with asymptotic variance $1/\Sigma 1/v_j$, i.e., to an estimate of 0.353 with a standard error of 0.068. The formal residual sum of squares, namely,

$$\text{RSS} = \Sigma T_j^2/v_j - (\Sigma T_j/v_j)^2/\Sigma 1/v_j \qquad (4.34)$$

gives a chi-squared test of homogeneity with $n_J - 1$ degrees of freedom. Here the value is 26.65 with 20 degrees of freedom. This is well short of statistical significance at conventional levels. Nevertheless, under a model with an additional between-study component of variance, τ_ξ, in the log odds, we have that

$$E(\text{RSS}) = (n_J - 1) + \tau_\xi\{\Sigma 1/v_j - (\Sigma 1/v_j^2)/\Sigma 1/v_j\}. \qquad (4.35)$$

This leads to an unbiased estimate in this case of 0.0259 corresponding to a standard deviation of 0.161. This is an estimate, albeit poorly determined, of the variation between studies in the 'true' odds ratio. If we regard the jth value as having a variance of $0.0259 + v_j$ we are led to a new weighted estimate of 0.381 with standard error 0.073.

There is little difference between the two estimates in this case, essentially because the relatively small studies do not differ at all systematically from the larger ones, which are more heavily weighted in the first estimate. The second estimate has a larger standard error.

Often the second analysis, or some equivalent to it, would be done only if there is statistically significant heterogeneity. It could, however, plausibly be argued that especially in case-control studies heterogeneity is to be anticipated and the null hypothesis of homogeneity is particularly implausible.

4.6 Maximum likelihood and REML

For the majority of the previous discussion we have motivated statistical procedures largely by relatively intuitive arguments essentially based on the generalized method of moments. In this estimating equations are formed by equating suitable functions of the data to their expectations under the assumed model. As we have noted, however, under normal theory assumptions and for balanced data the resulting analysis has a strong theoretical justification arising from its dependence on sufficient statistics. For unbalanced data, however, even under normal theory assumptions the reduction by sufficiency is at best partial and other arguments must be deployed.

The likelihood can be obtained quite generally under normal theory assumptions. If there were to be an evidence-based prior for the unknown parameters then Bayes's theorem would be available and the preferable route to interpretation. Failing that we may either use a flat prior in the hope that this will produce inference with good frequency properties or appeal to the method of maximum likelihood and its associated asymptotic properties. Often this will give satisfactory analyses but there is a difficulty whenever the number of parameters representing fixed features of the data is an appreciable fraction of the number of observations available. We illustrate this and an escape route by the simplest special case.

Suppose that in the balanced one-way arrangement with n_S observations on each of n_J groups the group means are regarded as individual parameters. i.e., not as random variables. That is, we consider the model

$$Y_{js} = \mu_j + \epsilon_{js}, \tag{4.36}$$

where the ϵ_{js} are independently normally distributed with zero mean and variance τ_ϵ. More generally suppose that the $n \times 1$ random vector Y has the form

$$Y = X\beta + \epsilon. \tag{4.37}$$

where now X is a $n \times d_\beta$ matrix of constants, β a $d_\beta \times 1$ vector of unknown fixed parameters and ϵ is an $n \times 1$ vector of independent normally distributed random variables of zero mean and variance τ.

Now for (4.37) the maximum likelihood estimates can be obtained by first maximizing the log likelihood

$$-(n/2)\log\tau - \Sigma(Y_{js} - \mu_j)^2/(2\tau) \tag{4.38}$$

over the μ_j for fixed τ to give the sample means $\bar{Y}_{j.}$ as maximum likelihood estimates. Then the resulting log likelihood is maximized over τ to give

$$\hat{\tau} = \Sigma(Y_{js} - \bar{Y}_{j.})^2/n: \tag{4.39}$$

for the general linear model the corresponding result is that

$$\hat{\tau} = \text{RSS}/n. \tag{4.40}$$

where the residual sum of squares RSS is defined in the usual way as $(Y - X\hat{\beta})^T(Y - X\hat{\beta})$, where $\hat{\beta}$ is the least squares estimate

$$\hat{\beta} = (X^TX)^{-1}X^TY. \tag{4.41}$$

It follows that in the two cases $E(\hat{\tau})$ is τ times

$$(n_S - 1)/n_S. \quad (n - d_\beta)/n, \tag{4.42}$$

respectively. For large $n - d_\beta$, $\hat{\tau}$ will be close to its expectation and thus unless n_S is large, or in the general case d_β is small compared with n. the estimate will be systematically less than τ. For example. if $n_S = 2$ and n large, the estimate will, with high probability, be close to one-half the parameter value.

This is an instance of a general phenomenon that if the number of nuisance parameters is large a maximum likelihood estimate may be very unsatisfactory.

In the present instance the resolution comes by dividing residual sums of squares by the appropriate degrees of freedom rather than by the total sample size. Formally this may be expressed as follows. In the one-way analysis we may apply an orthogonal transformation to each sample to replace the n_S values by $\bar{Y}_{j.}\sqrt{n_S}$ and $n_S - 1$ variables independently normally distributed with zero mean and variance τ. The contribution of the sample to the likelihood is thus the product of two factors, one depending on $\mu_j.\tau$ and one depending only on τ and involving the data only via $\Sigma(Y_{js} - \bar{Y}_{j.})^2$. In many circumstances, essentially when little is known initially about μ_j. the first factor contains little or no information about τ. Therefore for inference about τ we use only the second factor, leading to a log likelihood based on $\Sigma_j(n_S - 1) = n_J(n_S - 1)$ observations independently normally distributed with zero mean and variance τ. The corresponding maximum likelihood estimate has the 'correct' divisor, the degrees of freedom within groups.

In the more general case the orthogonal transformation of Y is to an internally orthogonalized form of the least-squares estimate $\hat{\beta}$ and to $n - d_\beta$ random variables independently normally distributed with zero mean and variance τ. Maximum likelihood applied to the latter alone leads to the usual estimate of τ, namely the residual mean square, the RSS divided by degrees of freedom for residual.

This procedure in which maximum likelihood is applied in a residual space is called REML, standing variously for reduced or residual maximum likelihood. In a broader setting it is a particular case of the use of marginal likelihood, i.e., likelihood in which selected features of the data are marginalized and ignored in calculating the likelihood. Note though that because of the special properties of the normal distribution the same estimates would have resulted had we conditioned on the values of the least-squares estimates and from this point of view the estimates are also based on conditional likelihood.

We now apply the same idea to the general model in which we write

$$Y = X\beta + AU, \tag{4.43}$$

where X, A are known matrices indicating the contributions of fixed and random effects, β is a vector of fixed parameters and U is a collection of random variables with a structure reflecting the formation of error via a combination of components. The matrix X is $n \times d_\beta$ and A is $n \times q$, where q is the total number of random variables needed to specify the model. In many applications it is possible to rewrite the defining model in the form

$$Y = X\beta + A_1 U_1 + \ldots + A_k U_k, \tag{4.44}$$

where the U_h are mutually independent vectors of independent components of size $q_h \times 1$ normally distributed with zero mean and variance τ_h, for $h = 1, \ldots, k$ and A_h is $n \times q_h$. Note that if the last component corresponds to the addition to each observation of a random term of constant variance then A_k is the identity matrix.

In general the covariance matrix of Y is $A^T \Sigma_{UU} A$, where Σ_{UU} is the covariance matrix of U, itself a function of variance components and in the more explicit form is

$$\Sigma_{YY} = \Sigma \tau_h C_h, \tag{4.45}$$

say.

Now the log density of the multivariate normal distribution is, except for a constant,

$$-\log \det(\Sigma_{YY})/2 - (Y - X\beta)^T \Sigma_{YY}^{-1} (Y - X\beta)/2 \tag{4.46}$$

from which maximum likelihood estimating equations can be obtained by differentiation with respect to the components of β and the variance components. To complete the argument it is in principle necessary to verify that the stationary value of the log likelihood is indeed a maximum and to introduce modifications if the local maximum is outside the parameter space, i.e., corresponds to negative estimated variance components.

For the corresponding discussion of REML estimates we transform from Y to $Y^* = RY$, where R is a symmetric matrix of rank $n - d_\beta$, namely

$$R = I - X(X^T X)^{-1} X^T, \tag{4.47}$$

this having a distribution not depending on β. The covariance matrix of the new vector is

$$\Sigma_{Y^* Y^*} = R \Sigma_{YY} R. \tag{4.48}$$

The argument now proceeds as before, where a generalized inverse has to be used for $\Sigma_{Y^* Y^*}$.

For further details, see Bibliographic notes. For balanced systems, the REML estimates are the ones introduced via the analysis of variance table; alternatively this can be proved directly by matrix algebra.

4.7 A different approach

We may motivate the methods set out here by a mixture of intuitive arguments derived from simple semi-descriptive decompositions based initially on balanced data and of likelihood arguments using normal-theory assumptions. The latter lead to formal optimality properties. In the balanced case these centre on the use of minimal sufficient statistics. The outcomes of this formulation are essentially point estimates of the relevant parameters supplemented by approximate likelihood-based confidence intervals for the parameters, these intervals being formally 'exact' in some special cases. Under moderate nonnormality the formal optimality properties are lost but the estimates usually retain appealing properties, for example of consistency, and at least in simple cases corrections to the confidence intervals for nonnormality can be made by modification of the formal degrees of freedom.

There is a different approach to these issues stemming in effect from a generalization of the second-order theory of the linear model, i.e., from the approach to least-squares theory based on unbiased estimates of minimum variance using only error assumptions of zero correlation and constant variance. In this second approach to estimation of components of variance only low moment assumptions are made about the component random variables and attention focuses on quadratic point estimates that satisfy conditions such as unbiasedness and minimum variance, or a criterion of minimum mean squared error. In many cases optimality can be made to hold only at a specified point of the parameter space, even though a condition such as unbiasedness holds universally. If the specified point is updated iteratively REML estimates typically result. The resulting estimates are named by a series of acronyms, especially MINQUE (minimum norm, quadratic unbiased estimators) and I-MINQUE (an iterative version).

4.8 Bibliographic notes

A detailed study of the ANZICS data in Example 4.1 is being conducted in collaboration with J. Moran and D. Firth. We are grateful to the Australian and New Zealand Intensive Care Society for permission to use the ANZICS data.

APACHE II (Knaus et al., 1985) is the most widely used algorithm for computing risk of death or probability of mortality in the literature on intensive care. A previous version, APACHE I, was found to be unsatisfactory. APACHE II was derived from a study of 5,815 patients from 13 hospitals but hospitals were not entered as fixed effects, and no adjustment for regionality was made.

There has been some discussion in the literature concerning problems with using a cut-point approach to modelling continuous explanatory variables. These include an increase in the Type I error and over-estimation of the effects at the defined levels (see, for example, Altman et al., 1994). In Example 4.1, a cut-point approach nevertheless provides a simple way of handling the length of stay variable, which has many very small values and an extremely large standard deviation.

The problem of complete separation effects in fitting logistic regression models via maximum likelihood is well known (Albert and Anderson, 1984). It was also discussed in the pattern recognition literature: see Duda and Hart (1973) and Ripley (1996). Firth (1993) removed the first-order term from the asymptotic bias of maximum likelihood estimates by penalising the likelihood by the Jeffreys invariant prior. This resolves the problem of separation effects in maximum likelihood estimation in binomial logistic models.

Olsen and Shafer (2001) developed a random effects model for use when the response variable has a binary and a continuous component, the former governed by a linear logistic regression equation and the latter by an ordinary regression equation. The random effects contributions to the two components are correlated.

The practice of combining estimates of the fundamental constants of physics by weighted least squares has a long history, although the values often quoted tend to be of the last and presumably most definitive determination. One of the first careful accounts of the issues involved in bringing together data from a number of studies, where the effects in different studies are similar but not in general identical is in an agricultural setting (Yates and Cochran, 1938). See also Cochran (1954) and for a brief review Cox (1982). Hedges and Olkin (1985) were influential in bringing these issues to the attention of social scientists. In medical statistics so-called overviews or meta-analyses are an integral part of evidence-based medicine.

The general matrix formulation of the unbalanced model is due to Hartley and Rao (1967). The use of likelihood formulations based on modified forms of the data goes back to Bartlett (1937). It was suggested in the context of balanced data by Anderson and Bancroft (1952) and developed in detail for unbalanced data by Patterson and Thompson (1971). A very careful account of the matrix algebra involved is given by Searle et al. (1992, Chapters 4 and 6). Marginal and conditional likelihoods were studied in generality by Kalbfleisch and Sprott (1970). For an account of asymptotic versions see, for example, Barndorff-Nielsen and Cox (1994).

The development based on MINQUE and similar criteria is described in detail by Rao and Kleffe (1988).

4.9 Computational/software notes

The books by Snijders and Bosker (1999). Venables and Ripley (1999) and Pinheiro and Bates (2000) contain suggestions and examples for fitting nonnormal and other models with random effects to unbalanced data.

Most of the widely available statistical packages (R. S-PLUS. SAS. Stata. etc.) include routines for fitting random effects models to unbalanced normal data via REML, for instance, via lme in R. Most packages also provide routines for fitting binary logistic regression models via maximum likelihood. as well as some routines for fitting generalized linear mixed models. The generalized additive model plots shown in Figure 4.2 were fitted using S-PLUS5; GAMs are available in most higher-level statistical packages.

The R code we used to obtain the (adjusted) maximum likelihood estimates of the regression coefficients and associated covariance matrices for the Australian intensive care data studied in Section 4.4 is set out in the Appendix (see comments in the Bibliographic notes on the bias-reduced logistic regression method utilized). Firth's R algorithm brlr for implementing the iterative estimation scheme is available from the CRAN website http://cran.r-project.org or. preferably. from a CRAN mirror site nearer to the reader's home. Alternatively. the bias in the maximum likelihood estimates and the separation effects could be adjusted to an extent using a simpler. crude empirical adjustment and maximizing the relevant likelihood directly.

Our experience suggests that the Stata program **gllamm** for generalized linear latent and mixed models (Rabe-Hesketh et al.. 2001. 2002) is valuable for fitting random effects logistic regression models of the form discussed in Example 4.1. but that it can be slow to run on a large dataset and complex models incorporating several correlated random effects. **gllamm** uses maximum likelihood and adaptive Gaussian quadrature to estimate the model parameters, and the program and manual are available from `http://www.iop.kcl.ac.uk/iop/departments/biocomp/programs/gllamm.html`. We are grateful to S. Rabe-Hesketh for advice on implementing **gllamm**, and for making a new version of the program available to us.

MLwiN can fit binary logistic regression and related models incorporating random effects, as can SAS's PROC NLMIXED procedure in version 8.1. Egret (Cytel software) fits logistic regression models with up to two normally distributed random effects. SAS NLMIXED and related routines use Gauss-Hermite quadrature for numerical integration and can handle non-normal response data with one or two random effects. GenStat contains a macro for Iteratively Reweighted-REML for hierarchical generalized linear models of which the mixed model form is a special case (Engel and Keen, 1994). Further comments are given in the Computational/software notes at the end of Chapter 5 on nonnormal models.

4.10 Further results and exercises

1. The analysis of the series of case-control studies has used methods based essentially on the approximate normality of the empirical log odds ratios from the separate 2×2 tables. Formulate the same issues directly in terms of the likelihood contributions from the separate tables and discuss how appropriate methods of analysis might proceed.

2. Obtain the log likelihood for the balanced one-way component of variance model by two routes. First regard the combined data as a $n_J n_S \times 1$ vector and note that its covariance matrix is block diagonal with elements of intraclass form. Hence invert the matrix and find its determinant. Then use the form of the multivariate normal density of Y to write down the likelihood and verify that it depends only on the overall mean and the sums of squares between and within groups which are thus sufficient statistics. Secondly proceed indirectly by first regarding the group means as fixed, applying an orthogonal transformation to each group to obtain the likelihood as a product of contributions from the $\bar{Y}_j$. and a term depending only on the sum of squares within groups. Then uncondition on the μ_j.

   Examine the extension to the balanced two-way and other more general balanced arrangements.

CHAPTER 5

Nonnormal problems

Preamble

Some of the corresponding problems for nonnormal distributions are addressed. First, some relatively simple issues concerning Poisson and binomial distributions (Sections 5.2, 5.3) are developed from first principles. In the Poisson case the latter involves in the simplest situation the negative binomial distribution. Then corresponding problems for survival or event history data are discussed (Section 5.4); here the random term is often called frailty. Finally, a much more general situation is outlined. Efficient methods in this case involve high-dimensional numerical integration and specialized software and these are briefly reviewed.

5.1 Preliminaries

The analyses discussed in previous chapters are based directly or indirectly on linear representations of continuous random variables and for many of their more formal properties depend on normality of the underlying distributions. Thus the analyses involving logistic regression are handled by first fitting separate models by maximum likelihood and then using essentially linear methods on the resulting estimates. We now discuss briefly some broadly parallel results centering on the Poisson and binomial distributions.

A key point implicit in the discussion is that because these distributions are defined by a single parameter the variance is a known function of the mean. In a sense this bypasses the need for one level of replication and implicitly defines an additional line in an analysis of variance table.

We then deal with broadly corresponding issues connected with survival data before proceeding to a rather general formulation.

5.2 Poisson distribution

5.2.1 A simple measurement problem

Most applications of the Poisson distribution are essentially concerned with counting point occurrences. In a Poisson process of rate ρ the number of points occurring in time t has a Poisson distribution of mean ρt. Sometimes, however, observation of the process of interest is contaminated by a background process of noise and in a sense this is broadly comparable to the addition of a random error of observation.

Suppose that in time t, the random variables U_t, V_t represent, respectively, the numbers of points in the process of interest and the noise process. The observed number is $Y_t = U_t + V_t$. We assume that V_t has a Poisson distribution of mean νt. In order to estimate ρ when ν is unknown we suppose that observations of the noise process alone are made for a separate time period t', giving V' points, say. Note that the notional parallel with the previous discussion could have been reinforced by writing

$$Y_t = \rho t + \nu t + \xi + \epsilon, \quad V'_{t'} = \nu t' + \epsilon'. \tag{5.1}$$

Now the estimate of ρ is $Y_t/t - V'_{t'}/t'$ with variance

$$(\rho + \nu)/t + \nu/t'. \tag{5.2}$$

Thus if observational effort per unit time is the same for the process of interest and for the background, the optimal ratio for t'/t is $\{\nu/(\rho+\nu)\}^{1/2}$.

This discussion assumes that signal and noise are independent and that there is no phenomenon such as counter blocking that would compromise either the independence or the Poisson assumptions.

5.2.2 A frailty model

A situation rather more parallel to the one-way analysis of Section 2.1 is as follows. We have counts $Y_1, \ldots, Y_n$ of randomly occurring events for n independent groups. It may be reasonable to treat Y_j as having a Poisson distribution with mean μ_j characteristic of the group in question. If now it is reasonable to treat group effects as random we write $\mu_j = \mu \xi_j$ where ξ_j is a random variable which we take to have expectation one in order to give the interpretation of μ as an overall mean. Sometimes ξ_j is called the frailty of the jth group; the terminology arises from applications in which the randomly occurring events are failures or adverse reactions of some kind.

Now if we denote a generic such random variable by Y we have that, because for a Poisson distribution the variance is equal to the mean,

$$
\begin{aligned}
E(Y) &= E_\xi E(Y \mid \xi) = E(\xi\mu) = \mu, & (5.3)\\
\mathrm{var}(Y) &= E_\xi\{\mathrm{var}(Y \mid \xi)\} + \mathrm{var}_\xi E(Y \mid \xi) = \mu + \mu^2\tau_\xi, & (5.4)
\end{aligned}
$$

where again $\tau_\xi = \mathrm{var}(\xi)$.

Thus from observations $Y_1, \ldots, Y_n$ the parameter μ can be estimated by $\bar{Y}$. and τ_ξ by

$$\{\Sigma(Y_j - \bar{Y}.)^2/(n-1) - \bar{Y}.\}/\bar{Y}.^2, \tag{5.5}$$

a formula broadly analogous to the formula for continuous variables, namely $(\mathrm{MS}_\xi - \mathrm{MS}_\epsilon)/n_S$.

More generally we have that the moment generating function of Y is

$$
\begin{aligned}
M_Y(p) &= E(e^{pY}) = E_\xi E(e^{pY} \mid \xi) = E_\xi\{\Sigma e^{pr} e^{-\mu\xi}(\mu\xi)^r/r!\} \\
&= E_\xi\{\exp(\mu p\xi - \mu\xi)\} = M_\xi(\mu e^p - \mu).
\end{aligned}
\tag{5.6}
$$

thus expressing the moment generating function of Y in terms of that of the frailty ξ. The probability generating function of Y can be obtained by direct transformation, writing $e^p = q$. where q is the argument of the probability generating function $E(q^Y)$.

From this, higher cumulants of Y can be obtained by expansion. For a fully parametric formulation the simplest results are achieved by assuming that ξ has a gamma distribution with unit mean. having thus the probability density

$$
\kappa(\kappa\xi)^{\kappa-1} e^{-\kappa\xi}/\Gamma(\kappa)
\tag{5.7}
$$

and moment generating function $(1+p/\kappa)^{-\kappa}$. It follows that the probability generating function of Y is

$$
\{1 + (\mu q - \mu)/\kappa\}^{-\kappa}.
\tag{5.8}
$$

a negative binomial distribution of mean μ and variance $\mu + \mu^2/\kappa$.

It is known that the second moment estimate of $\tau_\xi = 1/\kappa$ has a high efficiency relative to maximum likelihood via the negative binomial distribution provided μ is not too small. See Further results and exercises.

Alternative nonparametric approaches proceed either by equating higher order cumulants or by the use of empirical probability generating functions.

Another possibility which may occasionally be more appealing from a substantive point of view and in some ways empirically be quite flexible is to suppose that ξ has finite support. in particular support on just two points, say $\omega_0 < \omega_1$ with probabilities. respectively. $(\omega_1 - 1)/(\omega_1 - \omega_0)$ and $(1-\omega_0)/(\omega_1-\omega_0)$. Some formal simplification is achieved in the symmetrical version with $\omega_1 = 1/\omega_0 = \omega$. These are essentially what in some contexts would be called latent class models as contrasted with latent structure models in which ξ is assumed to have a continuous distribution. If for each j there is a vector of explanatory variables z_j possibly determining the mean, an initial log linear representation can be considered in the form $Y_j = \mu \exp(\beta^T z_j)\xi_j$, with the ξ_j as before in (5.7). and with β a vector of unknown regression coefficients. This will lead to a maximum likelihood analysis based on the negative binomial distribution. If. however. there is evidence that one of the regression coefficients has a random component the more complicated methods to be discussed later appear to be needed.

To a limited extent this argument can be extended to more complicated data structures. For example. a two-way classification with rows and columns both treated as random and without interaction on a multiplicative scale could be represented by supposing Y_{jv} to have conditionally a Poisson distribution of mean $\mu\xi_j\eta_v$. where the ξ_j and the η_v are independent random variables of unit mean and variances, respectively, τ_ξ and τ_η. The mean and covariance matrix of the Y_{jv} are specified by

$$E(Y_{jv}) \;=\; \mu, \tag{5.9}$$

$$\mathrm{var}(Y_{jv}) \;=\; \mu + (\tau_\xi + \tau_\eta + \tau_\xi \tau_\eta)\mu^2, \tag{5.10}$$

$$\mathrm{cov}(Y_{jv}, Y_{jw}) \;=\; \mu^2 \tau_\xi \; (v \neq w). \tag{5.11}$$

Estimation of the variance components is possible after equating the standard sums of squares of row and column means to their expectations, for example via

$$E\{\Sigma(\bar{Y}_{j.} - \bar{Y}_{..})^2/(n_J - 1)\} = \mu/n_V + \mu^2 \tau_\xi(1 + \tau_\eta/n_V). \tag{5.12}$$

Unfortunately the explicit formulae for marginal and joint distributions of the Y_{jv} resulting from gamma distributions for the ξ_j and η_v do not lend themselves to particularly simple likelihood based inference. Solution by Markov chain Monte Carlo would be feasible.

5.2.3 Underdispersion

The whole of the preceding analysis presupposes that while the Poisson distribution is a natural starting point for analysis and interpretation any departure from the Poisson distribution is in the form of additional random variation, in particular inducing a variance of each observation greater than the corresponding mean. That is, there is overdispersion relative to the Poisson distribution. It is, however, possible that there is underdispersion; this is shown by the variance being less than the mean. If the observations are counts in a stationary point process underdispersion means in particular that the process is not a Poisson process. Just one of many possibilities is that the process is a renewal process in which the distribution of intervals has coefficient of variation less than one. We shall not discuss this and other explanations further here.

5.2.4 An example

We now discuss an example involving a large amount of rather poor quality data where an emphasis on simple graphical methods aimed at hypothesis generation is appropriate. In some spheres this is called data mining.

Example 5.1. Associations between occupation and health. Carpenter et al. (1997) and later Law et al. (2001) use routinely collected national data from England to investigate associations between occupation and cancer in men aged 20 to 74 years over the period 1981 to 1987. Approximately half a million cancers were reported to the national cancer registration scheme in England during this period. Although national databases of this magnitude provide an extremely valuable resource for studying population trends in disease and health, they suffer limitations due to the incompleteness of individual patient information, potential confounding with lifestyle factors such as smoking, and their cross-sectional nature: in particular, occupational information is current only at the time of diagnosis and is often missing. So-called denominator data are also frequently not available, i.e., there is little or no knowledge of the numbers of individuals at risk in the different occupations. Moreover, it is not possible to link individual patient information so that multiple diagnoses within individuals will be missed.

A particular statistical difficulty arises in the interpretation of such data from the need to examine a large number of occupation by cancer-site associations simultaneously. The situation is that of considering a very large contingency table in which the rows represent occupations and the columns the specific cancers, and interest lies in the potential association implied in each cell, which in turn represents a particular occupation by cancer site combination.

The cancer registration data contain occupational information for 212 different job groups and 39 specific cancers, resulting in a 212×39 contingency table and 8,268 associations to be investigated. Consequently, a large number of false positives are to be expected within any formal testing framework, implying that there will be many apparent associations occurring by chance alone and not necessarily due to any meaningful association between occupation and cancer diagnosis. In an attempt to overcome this difficulty, Carpenter et al. proposed an empirical Bayes approach to produce shrinkage estimators of individual occupation-by-site combinations. The basis of the method is the assumption that the true notional values for the different cells are themselves observations from some underlying probability distribution, thus inducing a second level of variation over and above random Poisson counts.

The general idea is as follows. Let O_{js} be the observed number of cases in occupation j and cancer site s, $j = 1, \ldots, n_J; s = 1, \ldots, n_S$, and make the usual assumption that O_{js} has a Poisson distribution of mean $\mu_{js} = \exp(\nu_{js})$. Then assumptions about the formal structure of the μ_{js}, or equivalently the ν_{js}, are of interest, and a number of possible formulations arise. A sensible choice on which to base further study is

$$\nu_{js} = \rho_j + \gamma_s + \eta_{js}, \qquad (5.13)$$

where ρ_j is the overall (log) rate for each occupation, γ_s is the overall rate for each cancer site, and the η_{js} are perturbations specific to the occupation-site combinations. The η_{js} could be mostly the same, with interest focusing on any outliers representing potentially unusual combinations, or alternatively they may be generally dispersed about zero. Carpenter et al. found it necessary to regard the different occupations as having different variances, i.e., $\text{var}(\eta_{js}) = \tau_{\eta_j}$.

Write $O_{j.}$ and $O_{.s}$ for the occupation and site marginal totals respectively. Then E_{js}, the fitted number of cases, is derived under an overall model of proportionality as $E_{js} = (O_{j.}O_{.s})/O_{..}$. Then the log of the unadjusted *proportional registration ratio* (PRR) is

$$R_{js} = \log\left(\frac{O_{js} + 1/2}{E_{js} + 1/2}\right), \tag{5.14}$$

where the PRR is the ratio of the observed registrations in each cell to the number expected under simple proportionality. This represents the traditional approach to analysing data of this type. Note that the addition of $1/2$ avoids problems with the occasional cell with a zero count.

Under the model (5.13), R_{js} is an estimate of η_{js} with a variance that is reasonably estimated by

$$V_{js} = 1/(O_{js} + 3/2). \tag{5.15}$$

Clearly values obtained from very small observed counts will have high variance. The effect of the empirical Bayes approach is to smooth R_{js}'s based on small numbers of events more heavily than R_{js}'s based on large numbers of events. This then leads to the empirical Bayes or 'shrunk estimate' of R_{js},

$$R_{js}^* = \frac{R_{js}/V_{js}}{1/V_{js} + 1/\tau_{\eta_j}}. \tag{5.16}$$

To apply (5.16), an estimate of τ_{η_j} is needed. For occupation j, this can be obtained from the mean square of R_{js} minus the average of the V_{js}. The estimate can be negative, and if this is the case, it implies that for occupation j the variation across cancer sites is consistent with the simple Poisson model, i.e., random departures from the marginal distribution across all occupations, and is handled in the usual way by setting it to zero. Note that in epidemiology, R_{js}^* is interpreted as the log of an observed to expected ratio of site-specific cancer registrations, OER. Law et al. (2001) refer to $\tilde{\tau}_{\eta_j}$ as an *index of sensitivity* of occupation j.

The emphasis in analysis is to detect anomalous cells which may indicate abnormally high or low counts as modelled by the values of η_{js}. Since these are estimated by R_{js}^*, a sensible, informal way to proceed is to plot the ordered R_{js}^* within an occupation, say, versus their expected normal order statistics. The idea is to examine the plot for outliers, unusual patterns or features, or groups of points lying away from the overall smooth trend. In the present application, the first of these is the most likely and therefore of primary interest. The whole approach is one of generating hypotheses rather than testing them, and formal testing has been deliberately avoided so far in the analysis of these data.

Figure 5.1 shows the normal probability plot of the shrinkage estimates R_{js}^* (i.e., log OER) for cancer amongst male woodworking machinists. The

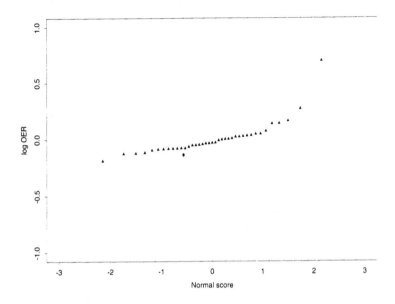

Figure 5.1 *Normal probability plot of the shrinkage estimates R^*_{js} (log OER) for cancer amongst male woodworking machinists. The five highest OERs. in decreasing order, are (highest) nose and nasal sinus. oesophagus. leukaemia. bone. Hodgkin's disease; the five lowest. in increasing order. are (lowest) skin other than melanoma, bladder, ill-defined and secondary. pleura. melanoma.*

data were kindly provided by Dr. N. Maconochie on the basis of work for Carpenter et al. (1997). The highest empirical Bayes estimate is for nose and nasal sinus cancer, and nasal cancer is a well-established occupational association for this employment group.

5.3 Binomial distribution

5.3.1 A parallel discussion

The account in the previous section for the Poisson distribution can be developed in a similar way for the binomial distribution. We shall sketch only a few points.

Suppose that $Y_1, \ldots, Y_n$ are independent variables. each specifying the number of positive outcomes in a group of m independent trials each with a binary outcome. The probability of a positive outcome in group j is denoted by ϕ_j so that Y_j has a binomial distribution with index m and parameter o_j. Now suppose that it is reasonable to treat the different groups as randomly sampled from some universe. so that o_j is treated as the value of a random variable denoted generically by Φ and having mean μ_ϕ and variance τ_ϕ. Then, by first arguing conditionally on $\Phi_1, \ldots, \Phi_n$ we have that analogously to (5.3) and (5.4)

$$E(Y_j) = m\mu_\phi, \tag{5.17}$$

$$\text{var}(Y_j) = m\mu_\phi(1 - \mu_\phi) + m(m-1)\tau_\phi. \tag{5.18}$$

From this μ_ϕ and τ_ϕ can be estimated by equating moments.

For a parametric version the formally simplest possibility is to take as the distribution of Φ the beta distribution with density $\phi^{\gamma-1}(1-\phi)^{\delta-1}/B(\gamma, \delta)$, where the normalising constant is the standard beta function. This has

$$\mu_\phi = \gamma/(\gamma + \delta), \tag{5.19}$$

$$\tau_\phi = \gamma\delta/\{(\gamma + \delta)^2(\gamma + \delta + 1)\}. \tag{5.20}$$

The probability distribution needed for a formal maximum likelihood analysis is the so-called beta-binomial distribution having

$$P(Y = r) = \frac{B(\gamma + r, m - r + \delta)B(r + 1, m - r + 1)}{(m+1)B(\gamma, \delta)}. \tag{5.21}$$

More complex problems are usually best studied by the methods for logistic regression discussed later.

5.3.2 Underdispersion

The representations given in the previous subsection all involve overdispersion relative to the binomial distribution. Underdispersion relative to the binomial distribution means that on replication the number of positive responses shows less variation than would be expected under a homogeneous model in which there is a constant probability throughout, combined with mutual independence. In analysis of variance terms, underdispersion is analogous to finding *less* variation in an upper mean square than would be expected by chance. The explanation is broadly the same too, namely the neglect of some feature within each group producing variation within a group that is not accounted for in the initial model.

In the present instance the simplest such model is to suppose that there is systematic variation of probability within each group, for example that the m trials within each group have probabilities $\phi_1, \ldots, \phi_m$, respectively, reproduced on replication across groups. By introducing an indicator variable for a positive response and regarding Y_j as the sum of the indicator variables, we have that

$$E(Y_j) = \Sigma\phi_s = m\bar{\phi}_., \tag{5.22}$$

$$\text{var}(Y_j) = m\bar{\phi}_.(1 - \bar{\phi}_.) - \Sigma(\phi_s - \bar{\phi}_.)^2, \tag{5.23}$$

showing underdispersion compared with the binomial distribution with parameter $\bar{\phi}_.$ unless all the ϕ_s are the same.

If the underdispersion induced by internal variation in the probabilities seems surprising, consider the special case when half the probabilities are zero and half one so that while $\bar{\phi}_. = 1/2$, the variance is zero.

5.4 Survival data

We now deal briefly with a special case of a random effect representation arising in connection with so-called failure or survival data. Here we observe for each of a number of individuals a random variable, T, that is the time from a suitable time origin to the occurrence of a well-defined point event, called here failure. For each individual we observe also a vector x of explanatory variables, assumed for simplicity fixed for each individual, i.e., we exclude so-called time-dependent explanatory variables. It is convenient to arrange that $x = 0$ corresponds to some standard condition that we call baseline. A very common complication in applications is right-censoring, i.e., it may be known only that the point event in question has not occurred by the end of the period of observation on that individual. Here, however, we are not particularly concerned with that aspect and for simplicity we assume that there is no censoring.

The objective is to study on the basis of data T_j, x_j for the jth individual how the failure time depends on the explanatory variables. There are two main formal models used in such discussions, the accelerated life model and the proportional hazards model. One way to define the accelerated life model is to suppose that there are notional failure times T_j, T_{j0} for the jth individual corresponding to the observed failure time and the failure time that same individual would have had had all their explanatory variables been at baseline. Then one supposes that

$$T_j = \psi(x_j)T_{j0}, \tag{5.24}$$

where often it is reasonable to take the acceleration factor in the form $\exp(-\gamma^T x_j)$. Here the T_{j0} are assumed independent and identically distributed random variables with a distribution, the baseline distribution, characteristic of the standardized baseline condition. There is an equivalent definition directly in terms of the survivor function $S(t) = P(T > t)$ avoiding the notional variable T_{j0}.

The proportional hazards model is defined via the hazard function, or age-specific failure rate, namely

$$h(t) = \lim P(t < T < t + \delta \mid t < T)/\delta, \tag{5.25}$$

where the limit is taken as $\delta \rightarrow 0+$. Under the proportional hazards model it is assumed that for the jth individual the hazard, shown now as depending on x_j, has the form

$$h_j(t) = \chi(x_j)h_0(t), \tag{5.26}$$

where $h_0(t)$ is the baseline hazard and often one would take the modifying factor in the form $\exp(\beta^T x_j)$. Note that both positive β and positive γ represent decreasing failure time with increasing x.

Now suppose that an additional level of random variability is introduced via a component e^{ξ_j} for individual j where by convention we can take $E(\xi_j) = 0$ or sometimes $E(e^{\xi_j}) = 1$. The variable ξ_j is usually called a frailty; see also Section 5.2. In the accelerated life model this changes T_{j0} to $T_{j0}e^{\xi_j}$. Thus if the form of the distribution of baseline failure time has been left unspecified the model is unchanged. If the distribution has been specified parametrically then typically the baseline parametric form will be changed but in a sense the form of dependence on x_j is unaltered.

The situation is in general slightly different for the proportional hazards model. Under the modified form of hazard function, the hazard of individual j is multiplied by a random term e^{ξ_j}. Then after marginalizing over the unobserved random variable, the hazard can be shown to be

$$\exp(\beta^T x_j)h_0(t)K'_\eta\{\exp(\beta^T x_j)\log S_0(t)\}, \tag{5.27}$$

where $K_\eta(.)$ is the cumulant generating function of the random variable $\eta = \log \xi$.

Except when η is constant, this is, except in very special cases, not of proportional hazards form. Now, while the proportional hazards assumption has some descriptive appeal, it is typically an empirical assumption and to be regarded as such. In effect the initial model contains a new random variable for each individual already and there seems no obvious need to introduce a second. Presence of a frailty term in a proportional hazards model can be detected only via a deviation from proportionality in the hazard functions; such departure would normally be better interpreted directly as evidence against proportionality, not necessarily via a frailty term. The accelerated model, by contrast, does have a direct physical interpretation and we have seen that a frailty term is in a sense nugatory in that context. In any case empirical experience suggests that introducing a frailty term into a proportional hazards analysis has little effect on the conclusions about the explanatory variables, except in extreme cases.

In the above discussion we have supposed that there is a single failure time observed for each individual. The situation is quite different if multiple point events are observed for each study individual. Here the incorporation of subject-specific frailty terms may be crucial. The simplest instance we observe for the jth individual is a Poisson process of point events of rate λ_j, say, so that the likelihood if r_j points are observed in the period of observation $t_j^{(0)}$ is

$$\lambda_j^{r_j}\exp(-\lambda_j t_j^{(0)}). \tag{5.28}$$

Now dependence on a vector of explanatory variables could provisionally be represented in the form

$$\lambda_j = \exp(\beta^T x_j)\lambda_0. \tag{5.29}$$

This supposes that all variation in the Poisson rate is accounted for by the explanatory variables and this would often be an unreasonable assumption. If we represent the additional variability by a random variable $e^{\xi_j} = \eta_j$. the likelihood contribution from individual j is

$$\int (\lambda_j \eta_j)^{r_j} \exp(-\lambda_j \eta_j t_j^{(0)}) dF(\eta_j). \tag{5.30}$$

In particular if we take the distribution of frailty in gamma form the integral can be evaluated in closed form as essentially the likelihood contribution from a negative binomial distribution. namely

$$\frac{\lambda_j^{r_j}}{(\kappa + t_j^{(0)}\lambda_j)^{r_j + \kappa}} \frac{\Gamma(r_j + \kappa)}{\Gamma(\kappa)}. \tag{5.31}$$

where κ is the index of the underlying gamma distribution of frailty.

The full likelihood of the data can now be found and maximum likelihood estimates of β, κ derived.

5.5 Some extensions

5.5.1 Preliminaries

The formulation in this book started with linear representations of balanced data. The extension to unbalanced data complicates analysis but not the essential ideas. Further extensions to nonlinear regression problems and to ones in which there is random variation in the regression coefficients and in which there may be particular forms of nonnormal distribution have been discussed above largely by relatively *ad hoc* modifications of the starting set-up.

We now consider in outline a rather general formulation in which. however, formally efficient methods of estimation involve high-dimensional integration. There are several routes to resolving the computational issues involved and the topic is an actively evolving one at the time of writing. The discussion given here is only introductory and use of the procedures will demand familiarity with one or more of the software packages currently available. with no doubt more to become available in the future. We favour wherever possible at least an initial analysis by the methods illustrated in earlier chapters and exemplified in the discussion of Example 4.1. This is not primarily because such methods are computationally simpler but rather that they are more transparent. That is. it is easier to see as the analysis develops where the main sources of variability lie.

In principle each new application may deserve a formulation from first principles. In practice, partly but not only for computational feasibility, it is desirable to have some broad families of models flexible enough to cover a range of applications. Two such important instances are normal theory nonlinear regressions and models based on generalized linear models, i.e., linear models formed via the exponential family of distributions.

An illustration of the former, namely the replacement of the linear structure by a nonlinear relation, is the change of a simple linear dependence of a response Y_j on an explanatory variable x_j to an exponential relation, such as

$$Y_j = \beta_0 + \beta_1 e^{-\beta_2 x_j} + \epsilon_j. \tag{5.32}$$

If the deviations ϵ_j remain represented by independent normal random variables of zero mean and constant variance τ_ϵ, we have a normal theory model of nonlinear regression. If now the data are divided into groups, corresponding for example to different study individuals, we may write the resulting observations Y_{js} in the above form with one or more of the $\beta_0, \beta_1, \beta_2$ depending on the suffix j. If that dependence is random we are led to a normal-theory nonlinear mixed effects model. Another possibility is that the upper level of random effect operates multiplicatively on the whole regression relation, leading to a representation of the form

$$Y_{js} = (\beta_0 + \beta_1 e^{-\beta_2 x_{js}}) e^{\xi_j} + \epsilon_{js}. \tag{5.33}$$

The second important general family depends on inserting additional random components into a generalized linear model, i.e., into a regression model with exponential family error structure. This includes as special cases not only the normal theory linear model but, in particular, broadly analogous systems with binary, binomial, Poisson or gamma error terms. These are the models whose analysis was encapsulated in the computer package GLIM and which are commonly referred to by the acronym GLM. If further random components are added we have the family of generalized linear mixed models, GLMMs.

Thus in disease mapping and small area statistics, we may need to account for overdispersion relative to, say, the Poisson distribution, and in addition represent spatial correlation. Correlated observations arise more generally in clustered and longitudinal data, for example, in individual repeated blood pressure measurements taken over the duration of a clinical trial. In fact in an earlier example, we established that data on blood pressure are log-normally distributed; moreover, introducing random effects into the linear predictor may be an efficient way to model directly observed correlation or overdispersion. Serial correlation may also be present and can be accommodated by some form of empirical stochastic smoothing such as is achieved by autoregressive errors.

Two valuable approximate ways of handling such problems are by the step-wise methods we have already illustrated and by transforming the response so that normal-theory methods are at least roughly applicable.

For example, suppose observed counts Y_{js} are approximately conditionally Poisson distributed. Then assuming multiplicative dependence on the explanatory variables, a quite general model for the transformed data would be

$$\tilde{Y}_{js} = \log(Y_{js} + 1/2) = x_{js}^T \beta + a_{js}^T \xi_j + \epsilon_{js} \tag{5.34}$$

where formally

$$\epsilon_{js} \sim N\left(0, \frac{\phi}{Y_{js} + 1/2}\right). \tag{5.35}$$

The x_{js} are explanatory variables for the fixed effects parameter β and the a_{js} are covariates for the random effects ξ_j. Here ϕ is a scale factor allowing for possible over- or underdispersion relative to the Poisson distribution, the latter corresponding to $\phi = 1$.

There are two rather different justifications for including the term $1/2$ in the definition of the empirical transform. One is as a kind of continuity correction to avoid singularities if there are occasional zero observations among the data. A more technical justification is that series expansion shows that if Y has a Poisson distribution of mean μ then for large μ

$$E\{\log(Y + 1/2)\} = \log \mu + O(1/\mu^2), \tag{5.36}$$

whereas for choices of the additive constant different from $1/2$ the error is $O(1/\mu)$. Now in principle it would be best to start from a representation of $\log \mu$ in terms of explanatory variables and then to find the transformation of the data that is closest to reproducing an analysis of the model for the parameters μ and, at least for unweighted analyses and for counts not too small, the above definition achieves that.

Similarly, for grouped binary data a natural choice of transformation is the logit probability of 'success', where again an appropriate normal theory model is fitted to the data.

In this second approach we apply an unbalanced weighted least squares-based analysis to the transformed data as contrasted with a maximum likelihood analysis directly to the originating model in the first approach. That is, in the second approach the assumed model is a linear approximation applying to the expected value of the empirically transformed observations rather than a model, for example of Poisson or binomial form, applied directly for the distribution of the observations themselves. While the two approaches are in a sense asymptotically equivalent and will often lead to essentially the same conclusions, the second approach has difficulties if, for example, there are supposedly Poisson distributed observations with an appreciable number of very small values, in particular zeros.

5.6 A more general formulation

The procedures outlined above are approximate in the sense that under the conditions specified the distributional properties of the estimates depend on approximations. There may also be loss of efficiency. Further, even when the procedures are in a sense asymptotically fully efficient and equivalent to maximum likelihood analyses of the model as specified, it is in general the case that methods based directly on the likelihood are preferable. In more technical language they are likely to have preferable properties in terms of higher-order asymptotic theory even when equivalent to the first order. Direct use of the likelihood is also more secure against irregular behaviour.

In fairly general terms we may start with a likelihood function that has the form

$$\text{lik}(\theta \mid \xi; y), \tag{5.37}$$

essentially the joint distribution of all the data given unknown parameters θ and initially conditionally on a vector of random terms ξ corresponding to random group, row, column, ... effects and to random contributions to regression coefficients. Typically but not necessarily ξ will be partitioned into sections, each section consisting of independent components identically distributed, each section with a different distribution.

The full likelihood is thus

$$\int \text{lik}(\theta \mid \xi; y) dF(\xi; \tau), \tag{5.38}$$

where $F(\xi; \tau)$ is the distribution function of ξ depending on parameters τ which are typically components of variance and generalizations thereof. In cases without time series and similar structure, ξ will consist of independent components so that $F(\xi; \tau)$ factorizes into a product component by component. The integral (5.38) will factorize into subintegrals but, even so, the dimension of each may be large.

5.7 Generalized linear mixed model

We now consider the special case where, given the random terms ξ, the observations have an exponential family distribution. Suppose that there are possibly several levels of random variability, the lower level corresponding to the exponential family distribution, for example the Poisson distribution, and the other to random variables ξ independently and not necessarily identically distributed. We suppose further the generalized linear form in which with observations Y_{js} with $j = 1, \ldots, n_J$ and $s = 1, \ldots, r_j$ there are explanatory variables and indicators x_{js}, a_{js} such that conditionally on the random effects

$$E(Y_{js} \mid \xi_j) \quad = \quad \mu_{js}(\xi_j) = h(x_{js}^T \beta + a_{js}^T \xi_j), \tag{5.39}$$

$$\text{var}(Y_{js}|\xi_j) \quad = \quad \phi b_{js} V(\mu_{js}). \tag{5.40}$$

where ϕ is a scale factor, the b_{js} are prior weights. typically associated sample sizes, h is the inverse link function associated with the exponential family and $V(\mu)$ is the function that for the exponential family in question relates the variance to the mean. Thus for the Poisson distribution $V(\mu) = \mu$ and $\phi = 1$.

Note that in this formulation there is a restriction in that the repeated observations within an individual or cluster are conditionally independent. so that if Y_j^* is the vector of the responses within the cluster then $\text{cov}(Y_j^*)$ is a diagonal matrix.

The random effects ξ_j are in the simplest formulation independently and identically normally distributed with zero mean and q-dimensional variance matrix $D(\tau)$, where τ is a vector of unknown variance components: D is often called the dispersion matrix.

The conditional independence of the observations within an individual or cluster allows us to write the exact marginal likelihood

$$\text{lik}(\beta. \tau: y) = \prod_{j=1}^{n_J} \int \prod_{s=1}^{r_j} f(y_{js}|\xi_j; \beta) g(\xi_j; \tau) d\xi_j. \tag{5.41}$$

where $g = h^{-1}$ is the so-called link function for the GLM. Then (5.41) is equal to

$$\prod_{j=1}^{n_J} |D(\tau)|^{-\frac{1}{2}} \int \exp\left\{-\frac{1}{2\phi} \sum_{s=1}^{r_j} d_{js}(y_{js}. \mu_{js}) - \frac{1}{2}\xi_j^T D^{-1}\xi_j\right\} d\xi_j. \tag{5.42}$$

where d_{js} denotes the deviance

$$d_{js}(y. \mu) = -2 \int_y^\mu \frac{y-u}{b_{js}V(u)} du. \tag{5.43}$$

The term in the exponent of (5.41) can be thought of as the log of a penalty attached to the likelihood for the generalized linear model.

5.8 Development of analysis

While the broad structure of the likelihood derived in the previous section is clear there are formidable difficulties in extracting useful information from it. There is one dimension of integration for each random component variable ξ_j and while special forms of the dispersion matrix D may well split the integrals into separate terms this will typically be an amelioration rather than a resolution of the difficulty.

The unknown parameters are the regression coefficients and the variance components. In principle there are several approaches to their study. One is to appeal to maximum likelihood, which is reasonable if there is an appreciable amount of information about all the parameters under estimation.

There are also two different approaches which may be called Bayesian. In one the parameters are assigned evidence-based priors representing external knowledge about the values of parameters likely on the basis of additional evidence. In the second, parameters, such as components of variance, are assigned relatively flat hyperpriors. Provided the dimensionality of this flat hyperprior is small procedures with good properties such as confidence or prediction intervals are likely to be achieved. The notion that flat priors represent initial ignorance or lack of information external to the data is tempting but is tenable at most in very low dimensional problems. Assignment of priors in several dimensions involves independence assumptions as well as ones of marginal form.

The issues of numerical integration are, however, more immediately challenging than those of the formal theory of inference. There are three ways to proceed. The most direct is direct or preferably adaptive quadrature. The word adaptive is used because it is important that the points at which the integrand is evaluated are adjusted to be appropriate for each subintegral.

The second approach, applicable when the integrals can be resolved into a sequence of one-dimensional integrals, is to use an analytical approximation, usually based on a few terms of a Laplace expansion. Such expansions are based on the idea that integrals involving an exponential of a function are dominated by behaviour of that function near its maximum.

The third method is Markov chain Monte Carlo (MCMC). In the Bayesian version of this a Markov chain is defined which has as its equilibrium distribution the posterior distributions of interest. The chain is then simulated a very large number of times and if the realizations appear to have converged to stationarity the frequency distribution of realized values, excluding a run-in period, is used to estimate the posterior distributions.

The second method can sometimes yield relatively simple interpretable results. Calculation of higher terms in the expansions may be feasible, especially if aided by computerized algebra. Higher terms are important to give at least a partial check on the adequacy of the approximations but there is often some uncertainty about the range of applicability of the approximations.

MCMC is a powerful and general technique. There is the possibility, in theory at least, that apparent convergence to a stationary state is illusory. Some protection can be achieved by starting the simulations from very different initial states. In the second Bayesian formulation described above there is the possibly much more serious issue as to whether the priors assigned are as innocuous as they may seem. Some protection can be gained by sensitivity analysis.

There are also at least two other approaches to these problems. Lee and Nelder (1996, 2001) study a notion of h-likelihood in which in effect realized values of individual random variables representing portions of variability are treated like unknown parameters. This is likely to be effective when there is substantial information about each such realized value. Another mode of analysis, called penalised quasi-likelihood, concentrates on the underlying estimating equations and their justification in a broader setting than a fully parametric one.

We have summarised above some of the key ideas involved. To proceed further it seems essential to become familiar with some of the specialized software involved, in particular to assess the strengths and weaknesses of the various approaches. This cannot easily be addressed in an introductory book such as this, quite apart from the rapid developments in progress alluded to above. Therefore we have confined further discussion to some brief bibliographic and computational notes and to an outline example. We think it likely that the longer term resolution will largely lie in direct numerical integration.

5.9 An outline example

To illustrate the generality of these formulations and the flexibility of the MCMC approach to their solution, we outline a particular application. For more details it is essential to consult the paper of De Angelis et al. (1998).

Example 5.1. Bayesian prediction of AIDS in England and Wales. A widely used method for predicting AIDS incidence is *back-calculation* or *back-projection*. The simplest form of the back-calculation model is represented by the convolution equation

$$a(t) = \int_0^t h(s)f(t-s)ds, \tag{5.44}$$

which relates the AIDS incidence $a(t)$ to the infection rate $h(s)$ and the density, $f(u)$, of the incubation period; $s = 0$ denotes the start of the epidemic which is usually taken to be 1981 in Australia and 1979 in the UK. Knowledge of any two of these functions allows the third to be determined. If a series of AIDS diagnoses $a(t)$ is observed, and the incubation density $f(u)$ is assumed known, then the past (unobserved) rate of HIV infection $h(s)$ can be estimated as a function of time; this is why the method is known as 'back-calculation'. Once the past infection curve has been reconstructed, (5.44) may be used again for projecting forwards in time to predict future AIDS incidence, typically for periods up to five years. New infections are assumed to arise according to a time-dependent Poisson process, so that AIDS incidence also follows a Poisson process with intensity $a(t)$.

De Angelis et al. (1998) use data from various sources on AIDS incidence. on sero-prevalence from surveys, on incubation times, on reporting delays and on the effect of treatment in postponing onset of AIDS in the later time periods. All this information has appreciable uncertainty and the combination produces a model with many sources of variability combining in a complicated way. It is striking that MCMC is able to produce estimates and predictions taking account of all these sources of uncertainty although of course various simplifying assumptions do have to be made.

5.10 Bibliographic notes

As noted in the text we give only a few key references to the very extensive literature on these topics.

Greenwood and Yule (1920) introduced the negative binomial distribution as a compound Poisson distribution in connection with an analysis of accidents to London bus drivers; Anscombe (1950) compared the theoretical properties of various methods of estimation of its parameters.

Law et al. (2001) re-analyse and further develop the work of Carpenter et al. (1997) on examining associations between occupation and health in men in England and Wales. Law et al. describe a simple approach taken to simulating statistical significance levels to aid interpretation of the normal probability plots, and further details and plots are to be found in both papers.

The discussion of over- and underdispersion associated with the binomial distribution goes back to the urn models of Lexis. For an early paper on the beta-binomial distribution, see Skellam (1948). Frailty in connection with survival data was introduced by Lancaster and Nickell (1980).

Random effect models associated with the normal-theory nonlinear regression model are discussed in detail by Pinheiro and Bates (2000). Their emphasis is on estimation through adaptive quadrature. Generalized linear models and generalized linear mixed models are described by McCullagh and Nelder (1989), Fahrmeir and Tutz (1994) and McCulloch and Searle (2001).

For a general account of Laplace expansions, see Barndorff-Nielsen and Cox (1988). Computerized algebra associated with such expansions is described by Andrews and Stafford (1999). For the inclusion of further terms in such expansions in the context of random effect models, see Solomon and Cox (1992), Breslow and Lin (1995) and Shun (1997), among others.

The use of Markov chains to perform numerical integration was suggested by Metropolis et al. (1953) with an important development by Hastings (1970). It has, however, come to prominence only much more recently in the light of modern computer developments. Green (2001) has given an excellent introduction. The book edited by Gilks et al. (1996) contains, in particular, numerous illustrative examples; Spiegelhalter et al. (2002) review methods and applications in medical research.

For penalized quasi-likelihood, see Laird (1978), Stiratelli et al. (1984). Breslow and Clayton (1993) and Breslow and Lin (1995).

Solomon and Wilson (2001) review the mathematical modelling and statistical procedures underlying methods used in industrialised countries for predicting future AIDS incidence and estimating past HIV infection rates. For the back-projection method, see Isham (1988, 1989) and Brookmeyer and Gail (1988).

5.11 Computational/software notes

The Computational notes in Chapter 4 are mostly relevant to fitting the models for nonnormal data described in this chapter.

The main statistical packages have some procedures for dealing with multilevel models. For details consult the detailed specification of R. S-PLUS, SAS, Stata, etc. R and S-PLUS offer. for example. `glmmGibbs` and `glmmPQL` for fitting generalized linear mixed models; the former estimates Bayesian generalized linear mixed models by Gibbs sampling (Myles and Clayton, 2001), and the latter utilizes penalized quasi-likelihood. Of the more specialized software we mention H. Goldstein's MLwiN and `gllamm` (Generalized linear latent and mixed models) (Rabe-Hesketh et al.. 2001) which runs in Stata. Rabe-Hesketh et al. (2002) provide a useful comparison of methods for estimation of generalized linear mixed models and of MQL. PQL, MCMC, Gaussian and adaptive Gaussian quadrature. respectively.

O. Christensen's R package `geoRglm` for generalized linear models with spatially correlated random effects is intended for geostatistical data. and is available from `http://www.maths.lancs.ac.uk/~christen/geoRglm/`.

Markov chain Monte Carlo methods are often implemented via some version of BUGS (Bayesian analysis using Gibbs sampling. Spiegelhalter et al., 1996), for example WinBUGS and versions for other systems. Linux and Unix in particular. Supplementary programming may be needed to deal with problems as complicated as the AIDS example sketched above.

5.12 Further results and exercises

1. Show that in a negative binomial distribution with known κ the maximum likelihood estimate of the mean μ is the sample mean. Suggest how this can be used to simplify maximum likelihood estimation of $(\mu. \kappa)$ when both parameters are unknown. Verify that it implies the orthogonality of the parameters.

   Obtain estimators of κ based on (a) the sample variance. (b) the proportion of zeros.

2. Show that if Y has a Poisson distribution then $\sqrt{Y}$. and to a better approximation $\sqrt{(Y + 3/8)}$, has approximately constant variance. Suggest how this could be used to test for overdispersion. Show that under a

model in which approximately $\sqrt{E(Y \mid \eta)} = \nu + \eta$, where $E(\eta) = 0$, the parameter $\text{var}(\eta)$ can be estimated. Examine the relation between this analysis and one based on the multiplicative model, showing them to be essentially equivalent when the observations are large.

3. Observations $Y_1, \ldots, Y_n$ are such that $E(Y_j) = \text{var}(Y_j) = \mu$, and that $\text{corr}(Y_j, Y_{j+h}) = \rho_h$. Obtain the conditions for large n under which the sequence has (a) overdispersion, (b) underdispersion. Note, however, that most more detailed models of serial correlation in count data induce failure of the variance-mean relation.

4. In the frailty model of Section 5.2.2 it is assumed that the random effect multiplies the underlying Poisson mean, μ. What would be the consequences if some other mode of combination, for example addition, were assumed and what data would be required for the differences between the two formulations to be detectable empirically?

5. Suggest further circumstances, additionally to that mentioned in Section 5.2.3, in which underdispersion relative to the Poisson or binomial distribution might be expected.

6. It is required to test whether a set of observations $Y_1, \ldots, Y_n$ is consistent with being a random sample from a Poisson distribution of unknown mean μ. Show how 'exact' tests can be obtained by examining the distribution conditionally on the mean $\bar{Y}$ and suggest suitable test statistics when it is suspected that the departure may be (a) overdispersion, (b) underdispersion, (c) serial correlation (d) an excess or deficiency of zeros. Compare the first two situations with use of standard chi-squared test for dispersion (Fisher, 1950).

CHAPTER 6

Model extensions and criticism

Preamble

A number of more special methods of analysis broadly connected with model criticism and improvement are discussed. These include the study of exceedances (Section 6.5), analysis of large numbers of small subsamples (Section 6.6), the use of nonlinear representations (Section 6.7), the systematic study of transformations of the response variable (Section 6.8) and the detailed study of the distributional form of the underlying random variables (Section 6.9).

6.1 Introduction

The previous discussion has been in terms of the simplest models relevant for the structures of data involved. Random terms are provisionally assumed independently distributed with zero mean and variances depending only on the stratum of error involved. Where formal tests and confidence intervals and considerations of optimality are concerned normality is typically assumed.

Now the very word *model* implies that an idealized representation of the real world is involved. It is unreasonable to expect that the model will represent faithfully all aspects of the system under study. The justification for starting with a model such as the ones above may be the quite strong one that it has been shown to apply reasonably well to previous similar situations through to the weaker one that it is the natural simplest starting point. In all cases, and especially in the second, some preliminary informal study of the data will be desirable.

We now turn to more extended discussion of model checking including the development of more elaborate models. Some such checking is essential if disasters are to be avoided. The kinds of procedure involved include

- relatively simple direct checks of the raw data,

- informal analyses and checks derived from the initial analysis,

- formal statistical tests of goodness of fit.

If the results of such analyses show evidence of model inadequacy the consequences may be

• an indication of major defects in the underlying formulation leading in extreme cases to a totally different approach to analysis;

• the isolation of unusual features of direct subject-matter interest;

• a suggestion of minor reformulation of the model;

• the pinpointing of probably defective observations or groups of observations. Often these are outlying single observations or groups of observations, but there are other possibilities, such as the reporting of identical values where supposedly independent replication is involved;

• some loss of precision in resulting estimates of key parameters and misspecification of confidence intervals. The latter often, but not necessarily, leads to underestimation of estimation error.

The role of tests of goodness of fit may be contrasted with less formal methods, graphical and tabular. Until some experience has been gained with the last two, there may be a temptation to overinterpret minor irregularities. Formal tests associated, say, with the graphical methods, essentially calibrate the plots against what is to be expected were the model in fact holding. Such calibration can also be achieved by simulation of synthetic data from the model under examination. The role of formal tests is also rather different depending on the amount of data, or more strictly the amount of effective information. With relatively small amounts of information apparently quite large anomalies can easily arise by chance and the formal tests provide some protection against overemphasis on the anomalies. With large amounts of data quite small departures from ideal behaviour are likely to be highly statistically significant. In both cases whenever clear evidence of a departure is found it will be for consideration on substantive grounds whether it is important enough to justify modification of the simplest analysis.

There are broadly two types of goodness of fit test. The so-called omnibus tests, often essentially chi-squared tests with fairly large degrees of freedom, give some power against a wide range of alternatives and hence have some ability to pick up unexpected kinds of departure. On the other hand the sensitivity against specific departures of particular interest is likely to be poor. We may contrast omnibus tests with those that are essentially equivalent to the fitting of an extended model with one or more additional parameters representing specific types of departure of subject-matter interest. On the whole we favour these, not only for their enhanced sensitivity, but also for their diagnostic power; they often point directly to specific modifications of the initial model.

Most of our discussion is for the simple one-way model. The more elaborate the structure the harder it is to make a sensitive analysis of the upper components of variation.

For the discussion in this chapter it is convenient to abandon our usual notation and instead to denote the number of groups by k and the number of replicate observations per group by r.

6.2 Modifications of structure

In carefully planned studies, especially the balanced types of data given in some of our examples, the form of the analysis of variance table is in principle determined by the way the data are obtained. in a randomized experiment by the randomization used. Which terms in the corresponding decomposition are to be regarded as random is then a subject-matter consideration. In broad terms the insertion of unnecessary components in the analysis of variance table may degrade precision but. unless very small numbers of degrees of freedom are involved. is not likely to be very misleading. On the other hand the omission of relevant terms may induce serious distortion. Such omission is likely to be the explanation of upper mean squares that are appreciably smaller than the appropriate error term. leading in particular to formal estimates of components of variance that are substantially negative.

A further possibility requiring a rather different analysis is that a particular sum of squares supposedly estimating a homogeneous source of variability is in fact a composite of degrees of freedom representing different sources of variability some large and some small. For example in a split-plot experiment the failure to separate the two levels of variability would induce this situation.

6.3 Outliers

One quite widely used approach to outliers. or more generally anomalous values, is to use methods that are insensitive to the occurrence of such values, so-called robust methods. Those methods that in effect involve replacing means and standard deviations as parameters describing location and scatter by other parameters such as medians and mean absolute deviations about the median are broadly inappropriate here. This is because these parameters do not retain the properties. in particular the additivity of variance as a parameter. that are in a sense the basis of our primary discussion.

We concentrate here on methods for detecting potential outliers. remarking of course that depending on context outliers may represent either aberrations best discarded or important indications of unexpected extreme behaviour.

In the one-way analysis outliers may affect individual observations within a group or may reflect outlying groups.

To check for possible outlying single observations there are a number of feasible procedures whose relative value depends on the number of replicate observations per group and on the number of groups.

With a small number of observations per group the most effective procedure is likely to be a probability plot of the sums of squares within the individual groups. In the perhaps unusual case where suspiciously small variances are of interest the plot should be of log variances. In this we order the variances or log variances and use as abscissa the expected order statistics of variances or log variances obtained from repeated samples of size r, say, from a normal distribution of unit variance.

For k samples the expected order statistics are for $j = 1, \ldots, k$ approximately at the $(j + 1/2)/(k + 1)$th quantiles of the corresponding theoretical distribution, namely the chi-squared or log chi-squared distributions with $r - 1$ degrees of freedom. A smooth nonlinear plot will be suggestive of an underlying nonnormal distribution of error whereas a small number of anomalous values at the upper extreme will be indicative of outliers. Anomalous extremely small values of variance may be an artificial consequence of rounding errors or of the mixing of heterogeneous sources of variability.

The above discussion is appropriate when there are a largish number of small groups, or more generally of small degrees of freedom. If, on the other hand, the individual sample sizes are not small it may be better to take residuals from the sample mean for each group and to plot against normal order statistics. In more complex systems inspection of residuals from, for example, a two-way additive representation may be the simplest starting point, although the intercorrelation of the residuals may make interpretation difficult. More easily interpreted although ultimately equivalent plots are obtained by taking the difference between each observation and a least-squares fit to the data omitting that observation.

To study possibly outlying groups we make a corresponding analysis of group means noting, however, that we are now in effect studying a convolution of error terms.

6.4 Robust estimation of an internal variance

Outlying observations are not to be unthinkingly discarded; they may be important indications of instability that need discussion and explanation. Sometimes, however, it is useful to regard any outlying observations as contamination of an underlying better behaved distribution whose variance can be taken as the target parameter. A way of estimating this uncontaminated variance that does not require explicit decisions about which are and which are not contaminating observations is as follows.

Suppose that observations are available on a large number of small groups and that the majority of the groups are unaffected by outliers. Then the probability plot of the ordered estimated variances should be linear over most of the range with a slope determined by the underlying uncontaminated variance. From the slope of the central part of the plot the required standard deviation can be estimated. A sensible compromise providing substantial protection against outliers without too much loss of efficiency in the absence of outliers is to use the central 80% of the distribution (Wilk and Gnanadesikan, 1964).

6.5 Model assessment: predicting exceedances

A reasonable component of variance model for a set of data should be capable of predicting the behaviour of other aspects of the system under study which are not directly described by the model. Qualitatively. the benefit of this is two-fold: firstly, it provides a means of assessing the goodness-of-fit of the model; and secondly, the model can be used to predict the behaviour of important features of the data using a small number of estimated parameters. The sensitivity of the predictions to changes in the parameters can also be studied.

In a number of fields, in particular many parts of biology. variables are characterised by their mean and variability and an important feature of data on such variables is sometimes the distributional behaviour of large values which exceed a given threshold level. which we call *exceedances*. Of course, if exceedances are the sole aspect of interest. one could study them directly and possibly nonparametrically. although this would give little or no insight into the determining parameters. for example. the relative importance of the variance components.

Assume that the data, possibly after suitable transformation. satisfy Situation 3 (equation (1.6) for between and within variation with replication) and let h be a given threshold level. We use estimates of the variance components τ_ξ and τ_ϵ to predict the behaviour of large data values occurring above h.

Define three parameters as follows: $E(N)$. the expected number N of exceedances; var(N), the variance of N: and $P(N = 0)$. the probability of observing no exceedances.

Note that conditionally on knowing the true group mean $\mu + \xi_j$ for group $j = 1, \ldots, k$, the observed number of exceedances N_j has a binomial distribution. In particular, the probability that Y_{js}, for $s = 1. \ldots . r$. is greater than h is

$$\Phi \left(\frac{\mu + \xi_j - h}{\sqrt{\tau_\epsilon}} \right). \tag{6.1}$$

where $\Phi(.)$ as usual denotes the standardised normal integral.

The unconditional parameters are derived as follows. Let $I_{js}(h)$ be the indicator function for the number of exceedances which is equal to one when $Y_{js} > h$, and 0 otherwise. Then the number of exceedances observed for the jth individual is the sum of r random variables

$$I_{j1} + \ldots + I_{jr} = N_j. \tag{6.2}$$

Because of the between group component ξ_j, the exceedances at different times are correlated. Write $\tau_. = \tau_\xi + \tau_\epsilon$. Then it is straightforward to show that for all j, s

$$E(I_{js}) = \Phi\left(\frac{\mu - h}{\sqrt{\tau_.}}\right), \tag{6.3}$$

$$\text{var}(I_{js}) = \Phi\left(\frac{\mu - h}{\sqrt{\tau_.}}\right)\left\{1 - \Phi\left(\frac{\mu - h}{\sqrt{\tau_.}}\right)\right\} \tag{6.4}$$

and

$$\begin{aligned}
E(I_{js}I_{jt}) &= P\left(\frac{\xi + \epsilon_{js}}{\sqrt{\tau_\epsilon}} > \frac{h - \mu}{\sqrt{\tau_.}}, \frac{\xi + \epsilon_{jt}}{\sqrt{\tau_\epsilon}} > \frac{h - \mu}{\sqrt{\tau_.}}\right) \\
&= \Phi_2\left(\frac{h - \mu}{\sqrt{\tau_.}}, \frac{h - \mu}{\sqrt{\tau_.}}; \frac{\tau_\xi}{\tau_.}\right),
\end{aligned} \tag{6.5}$$

where $\Phi_2(x, y; \rho)$ is the standardised bivariate normal distribution function of correlation ρ. Note that the I's are identically distributed with a common pairwise correlation coefficient ρ_I, say.

It follows that the expected value and variance respectively of N_j are

$$E(N_j) = r\Phi\left(\frac{\mu - h}{\sqrt{\tau_.}}\right), \tag{6.6}$$

$$\text{var}(N_j) = r\text{var}(I_{j1})\{1 + (r - 1)\rho_I\}, \tag{6.7}$$

with $r\text{var}(I_{j1})$ being the variance of N_j under the binomial assumption. Note that $\text{var}(N_j)/\{r\text{var}(I_{j1})\}$ is a measure of overdispersion relative to the binomial dispersion.

The unconditional probability of no exceedances is obtained from (6.1) by integration, and is

$$\int_{-\infty}^{\infty} \phi(x)\left\{\Phi\left(\frac{\gamma - x}{\zeta}\right)\right\}^r dx, \tag{6.8}$$

where $\gamma = (h - \mu)/\sqrt{\tau_\xi}$ and $\zeta = \sqrt{(\tau_\epsilon/\tau_\xi)}$.

Substituting maximum likelihood or analysis of variance estimates for the mean and the two variance component parameters, estimates of $E(N_j)$ and $\text{var}(N_j)$ can be found using standard statistical software or tables of the normal distribution; the integral (6.8) can be evaluated by numerical integration.

The fitted values of $E(N_j)$, $\text{var}(N_j)$ and $P(N = 0)$ are then compared with the analogous statistics calculated directly from the data, i.e.,

$\bar{N}_. = \Sigma_{j=1}^{k} N_j/k$, the observed average number of exceedances; $\Sigma_{j=1}^{k}(N_j - \bar{N}_.)^2/(k-1)$, the corresponding variance; and $\hat{P}_0$, the observed proportion of groups with no exceedances.

If the one-way nested model is a reasonable fit to the data, then the fitted values should reproduce the analogous statistics observed directly from the data.

Standard errors for the empirical estimates are available, to the extent that we have a single, roughly normal value for each group and normal theory single-sample methods are applicable. In calculating standard errors in the fitted values from the model, one could use either simulation or local linearisation, i.e., the so-called delta method, instead of numerical integration. However, there is a snag: if we want to compare observed and fitted values, errors in the N's and in the estimated variance components are correlated because they are obtained from the same data. Thus if we take the difference between the observed and fitted values, we can find the variance of each directly, but not the covariance, except by simulation or bootstrapping. However, it may frequently be reasonable in applications to calculate standard errors assuming independence and to regard the resulting standard errors of the differences as upper bounds.

6.6 Analysis of variability within small groups

6.6.1 Preliminaries

In many but not all the structures considered here there are at the lowest level groups each of small numbers of observations. Their role is partly to provide an estimate of the lowest error component of variance. In some cases, however, it may be worth making a more detailed analysis. For this we suppose that we have k independent samples of size r, leading to an initial analysis in which the samples are in effect replaced by means $\bar{Y}_j$ and estimates of variance s_j^2 with the implied representation that the initial observations Y_{jt} can be represented in the form $\mu_j + \epsilon_{jt}$. Because of the use of s as an estimate of standard deviation, we have used t for a running suffix in the current section.

The objectives of more detailed analysis include the following:

- the detection of outlying observations,

- the detailed examination of the distributional shape of the ϵ,

- the detection of nonconstancy of various kinds in $\mathrm{var}(\epsilon)$.

To some extent the first two have been studied in Sections 6.3 and 6.1, respectively. The emphasis in the present section is on procedures for small values of r. While the methods could formally be used on sets of means formed at a higher hierarchical level, the interpretation would not be straightforward because variation at such levels is formed from a convolution with lower levels of random variation.

Of course there has to be either some substantive interest in these issues, or they must be thought possibly to have serious impact on the primary analysis, for detailed study along the following lines to be sensible. In particular we assume that checking for rounding errors, digit preferences and stuck instruments has shown these aspects to be unimportant.

Suppose that the simplest model, namely that the individual s^2 correspond to independent estimates derived from normal distributions all with the same variance τ, fails. The main ways in which this can be represented are as follows:

- The ϵ_{jt} could be independently normally distributed with $\text{var}(\epsilon_{jt}) = \tau_j$, where

    - τ_j depends on some of the features in the original data configuration, for example on rows and/or columns in a two-way lay-out

    - τ_j depends on an additional whole group variable w_j

    - τ_j depends on μ_j, the corresponding expected value, often suggesting the need for a transformation of the underlying response variable

    - τ_j varies randomly from group to group

- The ϵ_{jt} are independently and identically distributed in a nonnormal distribution.

- The observations within a group have some structure which has been ignored, for example,

    - there may be a classifying feature which should be included in the analysis, this tending to make the s^2 too large as well as possibly distorting their distribution;

    - there may be serial correlation within each group.

These complications could occur in combination. While in principle it may be possible to distinguish empirically between some if not all of these possibilities, in practice it may be clear on general grounds which if any of these anomalies is both likely and, if it occurs, of substantive interest.

In many applications the most common values of r are likely to be two and three. Therefore we consider these cases first before outlining methods for larger values of r; clearly the larger values give much more scope for informative analysis.

6.6.2 Samples of size two

For samples of size two the data can be regarded as replaced by means and ranges, i.e., differences between the larger and the smaller observation in each pair. The ranges, which are proportional to the square roots of the estimated variances, are under the standard model distributed proportional to a chi-variable with one degree of freedom. i.e.. have a semi-normal distribution. This suggests a probability plot of the ordered ranges versus the expected values of the order statistics from the standardized semi-normal distribution.

Underdispersion in the plot. i.e.. the largest ranges being smaller than would be expected. suggests that the ϵ have a distribution with flatter tails than the normal, whereas overdispersion is consistent with several of the possible departures listed above. If at all possible any more detailed analysis should be guided by subject-matter considerations.

Various formal likelihood-based analyses are possible. Also note that for sampling a nonnormal distribution the even-order cumulants of the range are twice those of ϵ, enabling the latter to be estimated directly.

Under normal theory assumptions. the range and the sample mean are independently distributed. Therefore a plot of range versus sample mean should show no dependence. A simple correlation or regression test of independence can be used, regarded in principle as a permutation test. although in practice a normal-theory test will usually be satisfactory. Clear evidence of departure can then be explored. either by relating the range to the primary classifying relations of the data or by attempting to relate the underlying τ_j to μ_j via the corresponding sample means.

6.6.3 Samples of size three

With samples of size three. systematic skewness of the distribution can be examined directly, for example by plotting the order statistics of each sample against three equally spaced values on the abscissa. For a more formal treatment we temporarily abandon the first identifying suffix and write the order statistics of an arbitrary sample in the form $Y_{(1)} \leq Y_{(2)} \leq Y_{(3)}$. Define V by the equations

$$Y_{(3)} \quad = \quad Y_{(2)} + s \cos V + \sqrt{3} s \sin V. \tag{6.9}$$

$$Y_{(1)} \quad = \quad Y_{(2)} - s \cos V + \sqrt{3} s \sin V. \tag{6.10}$$

In the geometry of samples of size three from a normal distribution the spherical symmetry of the probability contours implies that s and V are independently distributed. Moreover it can be shown that V is uniformly distributed on $(-\pi/6, \pi/6)$. Thus departure from uniformity for V would be evidence of asymmetry. Under the model in which the underlying τ_j vary from sample to sample, i.e.. the normal theory overdispersed model. these distributional results are preserved.

6.6.4 More detailed study of log variance

Under the simplest assumptions the separate estimates of variance have the form

$$\tau X_d/d, \tag{6.11}$$

where X_d is a random variable with the standard chi-squared distribution with degrees of freedom $d = r - 1$ and we have omitted the suffix on $\tau = \sigma^2$. The separate groups are assumed to have independent realizations of X_d. It follows that the $\log s_j^2$ have essentially the log chi-squared distribution. Now the cumulant generating function of $\log s^2$ is

$$\log E(p \log X_d) + p \log \tau - p \log d =$$
$$p \log \tau - p + p \log 2 + \log\{\Gamma(p + d/2)/\Gamma(d/2)\}. \tag{6.12}$$

It follows that

$$E(\log s^2) \quad = \quad \log \tau - d - d \log 2 + \psi(d/2), \tag{6.13}$$
$$\mathrm{var}(\log s^2) \quad = \quad \psi'(d/2), \tag{6.14}$$

where $\psi(x)$ is the digamma function $d \log \Gamma(x)/dx$. An approximation to these formulae accurate enough for most purposes is that

$$E(\log s^2) \quad = \quad \log \tau - \{d - 1/3 + 1/(8d)\}^{-1}, \tag{6.15}$$
$$\mathrm{var}(\log s^2) \quad = \quad (2/d)\{1 - 1/(4d) + 1/(40d)\}^{-1}. \tag{6.16}$$

Except at $d = 1, 2$ these formulae can be simplified in an obvious way.

These results provide a sensitive and flexible if somewhat theoretically inefficient method of analysing collections of estimates of variance. Transform to log variances and then compute an analysis of variance of the transformed variables, in the light of whatever structure is relevant for the data. For example, if each cell of a row by column arrangement provided an estimate of variance by internal replication within the cell, the log transformed variances can be analysed by the two-way analysis: rows, columns and residual. To this analysis of variance can be added a theoretical line, i.e., with formally infinite degrees of freedom, with mean square the theoretical variance (6.14). Departures from the mean squares to be expected under the baseline model of constant error variance and normal distributions can thus be assessed.

Note that the correction for bias in $\log s^2$ as an estimate of $\log \tau$ is unnecessary if all the groups are of the same size r. If, however, it were decided to use the method when not all the group sizes are the same then a bias correction would be desirable. There might also be the need to take account of the differing precisions of the different estimates.

There is some loss of efficiency in the above analysis even when all the sample sizes are the same. This arises because least-squares-based methods are being applied when the random components have a log chi-squared distribution. A comparable likelihood-based analysis is possible based on a multiplicative model with gamma distributed errors of known index but is rather less transparent than the analysis of variance. The asymptotic efficiency of the latter relative to the likelihood-based approach is $2/\pi$ at $r = 3$ and rises slowly to one as d increases, the limit reflecting the asymptotic normality of the log chi-squared distribution. This implies that the direct analysis of log variances will often need supplementation if $r \leq 5$.

It can be shown that if the ϵ in the original model are independent and identically distributed with zero mean and fourth cumulant ratio γ_2 then the above results apply approximately with the variance of $\log s^2$ increased by the factor $1 + \gamma_2/2$. Thus if the analysis of variance of $\log s^2$ were to show all mean squares comparable but differing from the theoretical error then one, but only one, possible explanation would be nonnormality in the errors ϵ.

6.6.5 Autocorrelation

With samples sizes of three or more it is possible to examine the data for serial correlation between successive values as recorded in sequence. For samples of size three the lagged sum of products of successive values is

$$(Y_1 - \bar{Y})(Y_2 - \bar{Y}) + (Y_2 - \bar{Y})(Y_3 - \bar{Y}) = -(Y_2 - \bar{Y})^2. \tag{6.17}$$

That this is negative is essentially a consequence of the strong negative correlation between the residuals. A Helmert transformation shows that under a null hypothesis of independent normality with constant mean and variance, (6.17) and $Y_3 - Y_1$ are independently distributed with simple distributions. Two formal tests are possible.

The first essentially allows the variance to be different in each group, i.e., takes the overdispersion model as the null hypothesis. If

$$W = 3(Y_2 - \bar{Y})^2/\Sigma(Y_t - \bar{Y})^2 \tag{6.18}$$

the above results show that W has unit mean and variance $1/2$; moreover $\pi^{-1}\cos^{-1}(W - 1)$ has a uniform distribution, a result useful for graphical analysis. If now W is computed separately for each group the resulting mean $\bar{W}$ will have an asymptotically normal distribution of mean one and variance $(2k)^{-1}$.

The alternative approach takes as null hypothesis the full normal theory model, i.e., with constant variance. Then it is appreciably more sensitive to pool numerators and denominators and to use as test statistic

$$\tilde{W} = 3\Sigma(Y_{j2} - \bar{Y}_{j.})^2/\Sigma_{j,t}(Y_{jt} - \bar{Y}_{j.})^2. \tag{6.19}$$

It follows from the distributional results for the numerator and denominator that $\tilde{W}/(2 - \tilde{W})$ has the F distribution with (k, k) degrees of freedom. Approximately $\tilde{W}$ is normal with mean one and variance $1/k$.

6.7 Analysis by model elaboration: a nonlinear form

6.7.1 Types of nonlinearity

In any substantive application, a nonlinear model may be of intrinsic interest, or may be viewed simply as a convenient technical device for testing the adequacy of a standard linear formulation. Of course if the amount of nonlinearity is small, local linearization will produce a model of linear form, which will be typically unbalanced. We assume therefore the presence of at least some form of nonlinearity of the form described below which cannot be handled in this manner.

Our primary motivation here is to make suggestions, both formal and informal, for quite detailed analysis of departures from standard linear models. We do not attempt a careful classification of the types of nonlinearity that can arise in variance component models, but rather we describe some of the situations commonly encountered in practice. We begin with two linear models: the simplest one-way representation $Y_{js} = \mu + \xi_j + \epsilon_{js}$ and the most general form, $Y = x\beta + aU$, where β is a vector of unknown parameters defining the systematic part of the variation and the matrix a and vector U of random variables define the random structure; $j = 1, \ldots, k$ and $s = 1, \ldots, r$.

Nonlinearity can then arise in the following ways.

(i) μ or $x\beta$ is replaced by a nonlinear form typically applicable, for example, in modelling assay data.

(ii) The random components, for example ξ_j and ϵ_{js} in the one-way model, combine nonlinearly in some way.

(iii) The random and systematic parts of the model combine nonlinearly, for example, in an exponential growth model with random doubling time.

(iv) The normal theory based structure in the linear models is replaced by an analogous form for the exponential family. We described models of this type in the two preceding chapters.

6.7.2 A second-order model

The simplest model of type (ii) representing local departures from the standard normal theory linear formulation is

$$Y_{js} = \mu + \xi_j + \epsilon_{js} + \alpha_{20}\xi_j^2 + \alpha_{11}\xi_j\epsilon_{js} + \alpha_{02}\epsilon_{js}^2. \qquad (6.20)$$

Note that for interpretation in applications, it may be better to use the dimensionless forms for the α's, in particular, to replace α_{20}, α_{11} and α_{02} with $\alpha_{20}/\sqrt{\tau_\xi}$, $\alpha_{11}/\sqrt{(\tau_\xi\tau_\epsilon)}$ and $\alpha_{02}/\sqrt{\tau_\epsilon}$ respectively. For the purpose of our development, we retain the model as presented in (6.20).

This second-order representation models a quite general form of nonlinearity involving both skewness of the random effects and heterogeneity of the within-group variation. In particular, the model allows us to separate out three distinct features of interest in which the simplest one-way model may fail:

- nonnormality of the component ξ.
- nonnormality of the component ϵ. and
- a dependence between $\text{var}(\epsilon_{js}) = \tau_\epsilon$ and the associated ξ.

To interpret the α parameters. note first that we can write

$$\mu + \xi_j + \alpha_{20}\xi_j^2 = \mu' + \xi_j' \tag{6.21}$$

where $\mu' = \mu + \alpha_{20}\tau_\xi$ and ξ_j' is a nonnormal random variable of zero mean. approximate variance τ_ξ and third moment $9\alpha_{20}\tau_\xi^2$. Then the standardized third cumulant of ξ' is $\rho_3^\xi = 9\alpha_{20}\sigma_\xi$. and similarly for ρ_3^ϵ. The ρ's are dimensionless parameters and therefore more helpful for qualitative interpretation than the α's.

In practice, mild nonnormality in ϵ or possibly ξ may be of little concern. In some ways. the most interesting parameter is α_{11} which represents the possibility that the variance within a group may be correlated with the group mean. If there is a variance-mean relationship of this nature present in the data. standard models may lead to inappropriate predictions or supplementary analyses. Note that $\rho_{11} = \alpha_{11}\sqrt{\tau_\epsilon}$ is a dimensionless measure of the rate of change of the conditional standard deviation of $\text{var}(\epsilon_{js}) = \tau_\epsilon$ with $\mu + \xi_j$. Thus for example if ρ_{11} is moderate. say greater than 0.1. this indicates that an appreciable proportion of the variability within groups is attributable to fractional changes in the within-group standard deviation with the group mean. Such higher-order moment effects may have important substantive implications. but are usually ignored in analysis.

6.7.3 Statistical analysis

A simple graphical method for assessing α_{11} is to plot the log of the within-group sum of squares or standard deviation for the jth group against the group mean. A positive scatter corresponds to $\alpha_{11} > 0$. i.e.. the within-group variance increases with increasing group mean. and conversely for $\alpha_{11} < 0$. There are several ways of formally testing whether $\alpha_{11} = 0$. First, noting that under the null hypothesis $\alpha_{20} = \alpha_{11} = \alpha_{02} = 0$. we have that the quantities plotted are independent and an exact permutation test is available. One simple and effective procedure is to regress the log mean square within groups on the group j mean using least squares. Under normality. the residual mean square then has a known theoretical value and a permutation test could be applied. for example. if sensitive assessment of significance were an issue.

An alternative strategy is to attempt to represent the variance-mean relationship explicitly. One extension of the one-way model is to take

$$Y_{js} = \mu + \xi_j + e^{\kappa\xi_j}\epsilon_{js}. \tag{6.22}$$

where κ is a constant. This represents a linear dose-response relation with multiplicative error. A log linear dose-response with multiplicative error would be analysed on a log scale. The multiplicative term $e^{\kappa \xi_j}$ allows the within-group variation to change with ξ_j. When $\kappa = 0$, the standard linear normal theory model is obtained. Note that κ cannot be conveniently estimated from the ξ_j via maximum likelihood without further assumptions or external information. However, it is relatively straightforward to construct a score test for $\kappa = 0$, as the null hypothesis distributions are all that are required. If we consider the approximate linear form of (6.22), it is easy to show how κ is related to the interaction parameter α_{11}. In particular, taking

$$Y_{js} = \mu + \xi_j + (1 + \kappa \xi_j)\epsilon_{js}, \tag{6.23}$$

shows that $\kappa = \alpha_{11}/\sqrt{(\tau_\xi \tau_\epsilon)}$.

Illustration. Calibration studies of instrumental methods in analytical chemistry usually require a linear calibration curve to estimate the concentration of a component of interest in a sample from a given response. Two types of errors observed in many analyses are a proportional error term ξ that, even if always present, is noticeable only at relatively high concentrations, and an additive error term ϵ that also is present throughout but is noticeable mainly for small concentrations.

One possible model for the errors is to take

$$Y_{js} = (\mu + \beta x_j)e^{\xi_j} + \epsilon_{js}, \tag{6.24}$$

where Y_{js} is the response of the measuring apparatus, such as peak area, at concentration x_j, μ is the overall mean and β is a constant of proportionality; normality of the random terms is often a reasonable assumption. The model approximates a constant standard deviation for very low concentrations and a constant coefficient of variation for higher concentrations.

Of course one possible approach to statistical analysis based on the general model (6.20), especially for testing the null hypothesis $\alpha_{20} = \alpha_{11} = \alpha_{02} = 0$, is to use maximum likelihood. Here we pursue the less formal route of equating appropriate second-order and cubic statistics to their expectations and solving for the estimates of the α parameters. The cubic statistics are

$$S_{30} = \Sigma_{s,j}(\bar{Y}_{j.} - \bar{Y}_{..})^3, \tag{6.25}$$

$$S_{21} = \Sigma_{s,j}(\bar{Y}_{j.} - \bar{Y}_{..})^2(Y_{js} - \bar{Y}_{j.}), \tag{6.26}$$

$$S_{12} = \Sigma_{s,j}(\bar{Y}_{j.} - \bar{Y}_{..})(Y_{js} - \bar{Y}_{j.})^2, \tag{6.27}$$

$$S_{03} = \Sigma_{s,j}(Y_{js} - \bar{Y}_{j.})^3, \tag{6.28}$$

noting that S_{21} is identically zero.

Taking expectations of the other three cubic statistics under the nonlinear model (6.20) gives

$$E(S_{30}) = \frac{6r}{k}(k-1)(k-2)\alpha_{20}\tau_\xi^2 - \frac{6}{k}(k-1)\alpha_{11}\tau_\xi\tau_\epsilon$$
$$+ \frac{6}{rk}(k-1)(k-2)\alpha_{02}\tau_\epsilon^2, \qquad (6.29)$$

$$E(S_{12}) = 2(k-1)(r-1)\alpha_{11}\tau_\xi\tau_\epsilon + \frac{6}{r}(k-1)(r-1)\alpha_{02}\tau_\epsilon^2, \quad (6.30)$$

$$E(S_{03}) = \frac{6k}{r}(r-1)(r-2)\alpha_{02}\tau_\epsilon^2. \qquad (6.31)$$

If the variance components τ_ξ and τ_ϵ are known, then unbiased estimates of the α parameters can be found. Usually however τ_ξ and τ_ϵ are unknown. If we replace them by the standard estimates, the following are consistent estimates of α_{20}, α_{11} and α_{02}:

$$\tilde{\alpha}_{20} = \frac{k}{6\tilde{\tau}_\xi^2 r(k-1)(k-2)}\left\{ S_{30} + \frac{3}{k(r-1)}S_{12} \right.$$
$$\left. - \frac{(k-1)(k+1)}{k^2(r-1)(r-2)}S_{03} \right\}, \qquad (6.32)$$

$$\tilde{\alpha}_{11} = \frac{1}{2\tilde{\tau}_\xi\tilde{\tau}_\epsilon(k-1)(r-1)}\left\{ S_{12} - \frac{(k-1)}{k(r-2)}S_{03} \right\}, \qquad (6.33)$$

$$\tilde{\alpha}_{02} = \frac{r}{6\tilde{\tau}_\epsilon^2 k(r-1)(r-2)}S_{03}. \qquad (6.34)$$

These are then converted into estimates $\tilde{\rho}_3^\xi$, $\tilde{\rho}_{11}$ and $\tilde{\rho}_3^\epsilon$ of the dimensionless parameters using again the standard estimates of the variance components.

The sampling properties of these estimates can be developed fairly easily. In summary, under the null hypothesis of normality for large numbers of groups and repeated measurements within each group, the standard error of $\tilde{\rho}_3^\epsilon$ is $\sqrt{6}/(rk)$, and the corresponding standard error for $\tilde{\rho}_3^\xi$ is $\sqrt{(6/k)}$ if the estimate is essentially the standardized third cumulant of the k group means.

The above discussion for the one-way model can be readily extended to general balanced crossed and nested arrangements.

We have not specifically addressed the issue of detecting and handling kurtosis, although important kurtosis could potentially be accounted for by assuming Student t rather than normal errors. When the nonlinearity is of intrinsic and substantive interest, alternative models or methods of estimation may be appropriate and these are discussed in detail in Chapter 5.

6.8 Analysis by model elaboration: transformation

In some ways a more economical way of model checking, an improvement
as well as an extension of the range of application of linear methods, is
to suppose that some nonlinear transformation of the responses leads to
values more consistent with the linearity and with the normal distribution.

The need for transformation can be examined quite formally, as well as
informally via graphical and other methods. For example in the univariate
one-way model, scatterplots of the group mean versus the within-group
standard deviation over the range of plausible transformations provide a
simple way of determining whether transformation eliminates any variance-
mean relationships in the data. Histograms and quantile-quantile plots, for
example, can be examined for evidence of nonnormality or outliers, and in
some applications such informal analysis may be all that is required.

Example 6.1. The IPPPSH revisited. Figure 6.1 plots the patient standard deviations versus patient means for systolic blood pressure from the
IPPPSH (as described in Sections 2.1 and 3.2) over a range of possible
power transformations. Figure 6.1 indicates that the variance within patients is increasing moderately with increasing patient mean on the original
data scale. The relationship is virtually eliminated on both the log and reciprocal scales (the latter appears marginally preferable based on the univariate systolic data), noting also that there is an outlier with a very large
standard deviation but average mean value. Recall from Chapter 3 that
we discussed the fact that log transformation of systolic blood pressure
maximised the likelihood in the bivariate nested model incorporating components of covariance. Of course there are very compelling general reasons
for using a common transformation, providing a simplified interpretation,
for both blood pressure components.

A likelihood-based approach in which the transformation family is incorporated formally via one or more transformation parameters is in some
sense optimal.

Consider then a family of transformations indexed by λ. For given λ,
let $y^{(\lambda)}$ denote the transformed value of the data y. If for some λ the
$y^{(\lambda)}$ satisfy the balanced one-way nested model (1.6), the likelihood of the
original observations y is

$$(2\pi)^{-\frac{1}{2}kr}|\Omega|^{-1/2}\exp\left\{-\frac{1}{2}(y^{(\lambda)}-\mu)^T\Omega^{-1}(y^{(\lambda)}-\mu)\right\}J(\lambda;y), \quad (6.35)$$

where $J(\lambda;y) = \prod|dy_{js}^{(\lambda)}/dy_{js}|$ is the Jacobian of the transformation and
the product is taken over all observations; Ω is the covariance matrix. Direct
calculation shows that the determinant and inverse of Ω are given by

$$|\Omega| = \tau_\epsilon^{k(r-1)}(r\tau_\xi + \tau_\epsilon)^k, \quad (6.36)$$

$$\Omega^{-1} = -\tau_\xi\{\tau_\epsilon(r\tau_\xi + \tau_\epsilon)\}^{-1}J_\xi + \tau_\epsilon^{-1}I, \quad (6.37)$$

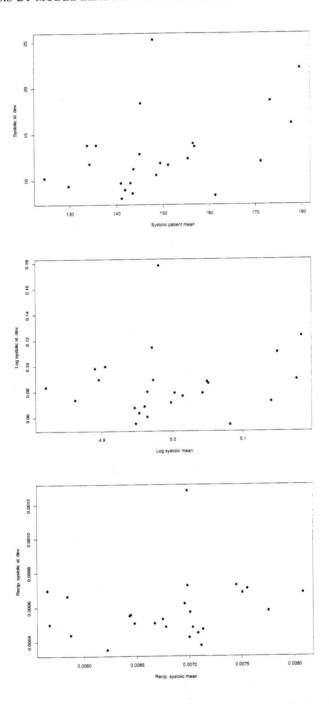

Figure 6.1 *Patient means versus standard deviations for systolic blood pressure for 25 men from the IPPPSH plotted for a range of power transformations: original data (top picture); log scale; reciprocal scale.*

where J_ξ has element one when rows and columns correspond to the same group, and I is the $kr \times kr$ identity matrix. This generalizes the formulation for the univariate one-way model.

Partitioning the exponent in (6.35) into its component sums of squares in the usual way leads to the log likelihood for the balanced one-way classification:

$$
\begin{aligned}
l(\lambda) \;=\; & -\frac{1}{2}kr\log(2\pi) - \frac{1}{2}k(r-1)\log\tau_\epsilon - \frac{1}{2}k\log(r\tau_\xi + \tau_\epsilon) \\
& -\frac{1}{2}[\tau_\epsilon^{-1}S_\epsilon(\lambda) + (r\tau_\xi + \tau_\epsilon)^{-1}\{rS_\xi(\lambda) + kr(\bar{y}_{..}^{(\lambda)} - \mu)^2\}] \\
& + \log J(\lambda; y),
\end{aligned}
\tag{6.38}
$$

where $S_\epsilon(\lambda)$ and $rS_\xi(\lambda)$ are the within- and between-group sums of squares, respectively, for the transformed data.

Thus in the balanced case the maximum likelihood estimates of μ, τ_ξ and τ_ϵ can be obtained by simply equating the canonical statistics to their expectations. For given λ, the maximum likelihood estimates are

$$
\hat{\mu}_\lambda \;=\; \bar{y}_{..}^{(\lambda)}, \tag{6.39}
$$

$$
\hat{\tau}_\epsilon^{(\lambda)} \;=\; \{k(r-1)\}^{-1}S_\epsilon(\lambda), \tag{6.40}
$$

$$
\hat{\tau}_\xi^{(\lambda)} \;=\; k^{-1}S_\xi(\lambda) - r^{-1}\hat{\tau}_\epsilon^{(\lambda)}. \tag{6.41}
$$

Alternatively, the log likelihood can be maximised directly, in which case λ can be estimated rather than assumed known, should this be desirable. However, it will frequently be the case in practice that precise determination of λ is not of prime importance as only the use of simple values of λ is sensible. Note that it is computationally desirable to use the normalised transformation $z^{(\lambda)} = y^{(\lambda)}/\{J(\lambda; y)\}^{1/(kr)}$ as this absorbs the Jacobian of the transformation to $y^{(\lambda)}$ into the component sums of squares, thereby inducing numerical stability in the variance component estimates for different values of λ.

6.8.1 Extensions to more complex models

In Example 3.1 we investigated the effects of log transformation on the IPPPSH data from the four-year cohort and fitted a bivariate hierarchical model with three nested effects. This is a special case of the balanced multivariate model with, in general, p nested effects whose log likelihood for the untransformed multivariate data takes the form

$$
\begin{aligned}
l(\lambda) \;=\; & -\frac{1}{2}pkr\log(2\pi) - \frac{1}{2}k(r-1)\log|\Omega_\epsilon| - \frac{1}{2}k\log|r\Omega_\xi + \Omega_\epsilon| \\
& -\frac{1}{2}[\operatorname{tr}\{\Omega_\epsilon^{-1}S_\epsilon(\lambda)\} + r\operatorname{tr}\{(r\Omega_\xi + \Omega_\epsilon)^{-1}S_\xi(\lambda)\} \\
& + kr(\bar{y}^{(\lambda)} - \mu)^T(r\Omega_\xi + \Omega_\epsilon)^{-1}(\bar{y}^{(\lambda)} - \mu)] \\
& + \log J(\lambda; y),
\end{aligned}
\tag{6.42}
$$

in which the determinant and inverse of Ω are given by

$$|\Omega| = |\Omega_\epsilon|^{k(r-1)}|r\Omega_\xi + \Omega_\epsilon|^k, \tag{6.43}$$

$$\Omega^{-1} = -\Omega_\xi\{\Omega_\epsilon(r\Omega_\xi + \Omega_\epsilon)\}^{-1}J_\xi + \Omega_\epsilon^{-1}I. \tag{6.44}$$

For the bivariate blood pressure model $p = 2$ and we use the power transformation family $y^{(\lambda)} = (y^\lambda - 1)/\lambda$ for $\lambda \neq 0$ and $y^{(\lambda)} = \log y$ for $\lambda = 0$.

The transformation theory generalizes directly to any component of variance model, although for unbalanced models the estimates need to be found numerically.

6.8.2 A special nonlinear form and parameterisation

One useful transformation model assumes, in effect, that for some κ the data satisfy the model

$$Y_{js}^{1/\kappa} = \mu + \xi_j + \epsilon_{js}, \tag{6.45}$$

where the right-hand side is the normal theory representation. Provided that μ is very much greater than $\sqrt{(\tau_\xi + \tau_\epsilon)}$, we obtain the following expansion for Y up to quadratic terms in the random effects:

$$Y_{js} = \mu^\kappa + \kappa\mu^{\kappa-1}\xi_j + \kappa\mu^{\kappa-1}\epsilon_{js} + \frac{\kappa}{2}(1-\kappa)\mu^{\kappa-2}(\xi_j^2 + 2\xi_j\epsilon_{js} + \epsilon_{js}^2) \tag{6.46}$$

which is of the general nonlinear form (6.20) with the restriction that, to within the scaling factors employed, $\alpha_{20} = \alpha_{02} = \alpha_{11}/2$.

In practice it is important that some effort should be expended on investigating suitable transformations of the data to induce additivity of the effects and/or desirable normality properties, regardless of whether the components of variance are of interest for their own sake or used primarily as a device for increasing the precision of other contrasts of interest. It is often the case, however, that the choice of 'best' transformation may not be clear. For instance, a transformation may adjust some features of the data to the standard assumptions, such as eliminating variance-mean relationships, but might not achieve normality of the error distributions. It is therefore of interest to know that there are parameters whose estimates remain stable under transformation. In particular, the intraclass correlation coefficient ρ is approximately orthogonal to changes in the scale of measurement. The intraclass correlation, or equivalently the ratio of variance components, is often of substantial interest in its own right, for example, in genetics, it measures the proportion of an effect that could be attributed to heritability.

In applications of variance component models, it is common to have fixed as well as random effects in which the overall mean μ in the usual one-way model is replaced by a linear predictor term $\Sigma_l x_{jl}\beta_l$, where the β's are unknown regression coefficients to be estimated. Then it can be shown that the *coefficient ratios* β_i/β_l also remain relatively stable to changes in the scale of measurement. This is an example of an important general result that ratios of regression coefficients are interpretable quantities measuring the substitutability of one covariate for another, and which can be robustly estimated without being too dependent on the specific assumptions in the model. We discuss this and further results in the Bibliographic notes, but remark here that even in situations where the covariance structure is not the primary focus of analysis, approximate orthogonal parameterisation may help simplify the analytical and numerical procedures.

An alternative parameterisation to the above is to consider what functions, if any, of the variance components τ_ξ and τ_ϵ in the one-way model are orthogonal to ρ. For ease of notation, let $\theta = \tau_\xi + \tau_\epsilon$. Following Solomon and Taylor (1999), we seek a reparameterisation from (μ, θ, ρ) to parameters (ψ_0, ψ_1, ρ) such that ψ_0 and ψ_1 are orthogonal to ρ. The new parameters are formulated as convenient functions of the arbitrary constants of integration involved in solving the relevant differential equations for orthogonality. Clearly μ is already orthogonal to ρ so $\psi_0 = \mu$.

Taking l to be the log likelihood for the one-way model, in general unbalanced, then for orthogonality we want the expected value of the relevant inner product to be zero, i.e.,

$$E\left(\frac{\partial l(\theta, \rho)}{\partial \rho}\frac{\partial l}{\partial \theta}\right) = 0. \tag{6.47}$$

In other words, we want the gradient of the log likelihood to remain orthogonal to the likelihood planes as ρ changes. The differential equation for θ specified by (6.47) is

$$i_{\theta\theta}\frac{\partial\theta}{\partial\rho} = -i_{\theta\rho}, \tag{6.48}$$

where $i_{\alpha\beta}$ denotes an element of $-E\{\partial^2 \log L/\partial\alpha\partial\beta\}$ which are here given by

$$i_{\theta\theta} = \frac{r_.}{2\theta^2}, \tag{6.49}$$

$$i_{\theta\rho} = \frac{1}{2\theta}\sum_j \frac{\rho(1-r_j)r_j}{(1-\rho)(1-\rho+\rho r_j)}. \tag{6.50}$$

Substituting the information terms we obtain

$$\frac{1}{\theta}\frac{\partial\theta}{\partial\rho} = \frac{1}{r_.}\sum_j \frac{\rho(r_j-1)r_j}{(1-\rho)(1-\rho+\rho r_j)}. \tag{6.51}$$

The expression on the right-hand side of (6.51) can be integrated exactly

Table 6.1 *Estimates of the α parameters and their dimensionless equivalents ρ for the IPPPSH blood pressure data. D: diastolic; S: systolic.*

	$\tilde{\alpha}_{20}$	$\tilde{\alpha}_{02}$	$\tilde{\alpha}_{11}$	$\tilde{\rho}_3^\xi$	$\tilde{\rho}_3^\epsilon$	$\tilde{\rho}_{11}$
D	0.0166	0.0114	0.0248	0.7565	0.7150	0.1254
$\sqrt{D}$	0.2851	0.0924	0.2908	0.6686	0.3113	0.0758
$\log D$	1.1656	0.0410	0.7568	0.5703	0.0276	0.0411
S	0.0092	0.0043	0.0121	1.1428	0.5026	0.1669
$\sqrt{S}$	0.1915	0.0455	0.1773	0.9587	0.2251	0.0986
$\log S$	0.9409	0.0151	0.5308	0.7710	0.0117	0.0479

to give the solution for θ as

$$\psi_1 \Pi_{j=1}^k [(1-\rho)^{-(r_j-1)}(r_j-1)^{-(r_j/2-1)}\{1+(r_j-1)\rho\}^{-1}]^{1/r}, \quad (6.52)$$

where ψ_1 is an arbitrary constant. Rearranging this and dropping terms which depend only on r_j, we obtain the parameter orthogonal to ρ:

$$\psi_1 = \tau_\epsilon \left\{ \prod_{j=1}^k \left(\frac{\tau_\epsilon + r_j \tau_\xi}{\tau_\epsilon} \right)^{1/r.} \right\}. \quad (6.53)$$

In the balanced case ψ_1 reduces to $\tau_\epsilon^{1-(1/r)}(\tau_\epsilon + r\tau_\xi)^{1/r}$, which can be thought of as a form of the geometric mean of the two variances τ_ϵ and $\tau_\epsilon + r\tau_\xi$.

Example 6.2. An analysis of nonlinear form. To illustrate some of the above discussion, we return to the IPPPSH blood pressure data. To simplify the analysis, we have averaged the replicate measurements at each patient visit, and appropriately combined the between-replicate and between-visits variance components estimated from the bivariate nested model (3.18) and (3.19).

Table 6.1 sets out the α and ρ statistics for diastolic and systolic blood pressure analysed separately over a range of suitable transformations.

The results are consistent with the log transformation model under which we expect $\alpha_{20} = \alpha_{02} = \alpha_{11}/2$ and observe 0.0166, 0.0114, 0.0248 for diastolic blood pressure, and 0.0092, 0.0043, 0.0121 for systolic, demonstrating reasonable agreement. The transformation model can be regarded as being really quite restrictive compared to the more general second-order model (6.20) but is clearly appropriate here.

The results presented in Table 6.1 do more than confirm the application of the transformation model though. Recall that ρ_3^ξ and ρ_3^ϵ represent the degree of skewness of the between-patient and within-patient random variables, respectively. For diastolic pressure, the patient and visit effects are of comparable but modest skewness on the original scale, whereas the distribution of the between-patients effects for systolic pressure has a much longer tail. These differences are most likely to be due to the target control level applied to diastolic but not systolic blood pressure. It is interesting that log transformation virtually eliminates the within-patient skewness and reduces that between patients.

Perhaps the parameter of most interest is ρ_{11}, which gives the rate of change in the standard deviation of the within-patient variation for changes in the patient mean. This rate is 12.5% for diastolic and 16.7% for systolic blood pressure, which indicate that a moderate proportion of the within-patient variation is attributable to the fractional change of the standard deviation within patients. Again, the power transformations reduce these dependencies in the data which are virtually eliminated on the log scale.

The impact of higher-order moment quantities such as the ρ's on the analysis and interpretation of data can be difficult to assess. In this particular application, we can speculate that the variability detected in the higher-order moments may well be important biologically. For example, published standards for treatment dosages for hypertension typically assume that the relationship between the 'true' patient mean and the 'true' within-patient variation is known, and that the conditional standard deviation of the within-patient variation does not change with changes in the patient mean. Whilst this is approximately true for the data on the log scale, published dosage tables are usually based on blood pressure data measured on the original scale (mmHg), and this analysis illustrates that too superficial an understanding of the underlying relationships in data on blood pressure might well lead to inappropriate treatment.

6.9 Nonparametric estimation of distributional form

While very detailed assessment of distributional form is often not needed, we consider in outline how it might be achieved, concentrating on the one-way lay-out with therefore two distributions to be estimated that of the lower component ϵ and that of ξ, the upper component.

Suppose then that we have k groups each of r observations and that we assume that the means of the groups are arbitrary and that we work with the usual representation

$$Y_{js} = \mu + \xi_j + \epsilon_{js}, \tag{6.54}$$

where now the ξ_j and the ϵ_{js} are independent sets of mutually independent random variables with distribution functions, etc. denoted by $F_\xi(.), F_\epsilon(.)$. Note that in this model while the condition that ξ_j and ϵ_{js} are uncorrelated is a matter of definition the mutual independence is a quite strong assumption, essentially that the amount of internal variation is independent of the conditional mean.

An approach to the estimation of the two distribution functions that in principle works for arbitrary distribution functions proceeds by relating the cumulant generating functions of linear combinations of the data to the underlying cumulant generating functions, $K_\xi(.), K_\epsilon(.)$. These equations can be solved for the underlying functions. Then the cumulant generating functions of the data are replaced by empirical cumulant generating functions and the resulting moment generating functions inverted by numerical Fourier transformation to obtain the required estimates of $K_\xi(.). K_\epsilon(.)$. This method is available even for $r = 2$ but requires very large k to be effective. See the Bibliographic notes.

A more restrictive approach is to examine third and perhaps fourth order cumulants. This is equivalent to using the local behaviour of the cumulant generating functions near the origin.

The most direct approach is to note that the residuals, for example. $Y_{js} - \bar{Y}_{j.}$, have the form

$$\epsilon_{js}(1 - 1/r) + \Sigma_{t \neq s}\epsilon_{jt}/r. \tag{6.55}$$

The expected value of the cube of this is

$$(r - 1)(r - 2)\kappa_{\epsilon,3}/r^3. \tag{6.56}$$

where $\kappa_{\epsilon,3}$ is the third cumulant of the distribution F_ϵ. This leads to an unbiased estimate of $\kappa_{\epsilon,3}$ based on the sum of the cubes of the residuals.

Similarly we may consider $(\bar{Y}_{j.} - \bar{Y}_{..})^3$ which has expectation

$$(k - 1)(k - 2)\kappa_{\xi,3}/k^3 + (k - 1)(k^2 - 2k + 2)\kappa_{\epsilon,3}/(k^3 r^2). \tag{6.57}$$

Thus once the skewness of ϵ has been estimated, or possibly assumed known, the skewness of ξ can be estimated.

Now the first part of this procedure. namely that for the estimation of $\kappa_{\epsilon,3}$, fails if $r = 2$; this is clear in that the corresponding residuals are differences of pairs of identically distributed random variables and hence are symmetrically distributed. The possibility of paired observations is important in applications, and maybe even the most common special case in many fields. In that case we may proceed slightly differently basing the estimation on the two cubic expressions

$$Y_{j1}Y_{j2}(Y_{j1} + Y_{j2})/2 \text{ and } (Y_{j1} + Y_{j2})(Y_{j1} - Y_{j2})^2)/2 \tag{6.58}$$

with expectations, respectively, $\kappa_{\xi,3}, \kappa_{\epsilon,3}$.

In fact for $r \geq 3$ the appropriately symmetrical cubic functions of the observations for estimating the skewness components are not unique and in principle at least an asymptotically efficient combination can be found.

For fuller details, see the papers listed in Bibliographic notes.

6.10 Bibliographic notes

The development on exceedances in Section 6.5 follows Solomon (1989) in which an application to the IPPPSH blood pressure data can be found. Cox and Solomon (1986, 1988) present formal and informal methods for the analysis of variability and serial correlation in a large number of small samples of observations. Solomon and Cox (1992) outline how to apply the nonlinear form (6.20) to hierarchical models with three nested effects.

Score tests for homogeneity of variance in random effects models have been proposed by Commenges and Jacqmin-Gadda (1997). Score tests based on the extended one-way model (6.22) have been investigated by Hunt and Solomon (1999). Rocke and Lorenzato (1995) describe models similar to (6.24) and extend their formulations to measuring gene expression from microarray data (Rocke and Durbin, 2001). Calibration models of this type have also been applied in environmental science (see for instance Zorn et al., 1997, 1999). An analysis of transformations for models with components of variance and covariance is given in Solomon (1985).

The original idea of orthogonality of parameters goes back to Hurzurbazar (1950) for which Cox and Reid (1987) introduced the insensitivity interpretation. Taylor et al. (1996) proved the approximate orthogonality results for the case $\psi = 0$ in Section 6.8.1, i.e., for log transformation, which was then generalized by Solomon and Taylor (1999) to general transformations for models with variance components; the idea is to orthogonalise relative to the transformation parameter. Solomon and Taylor suggest that a natural way to parameterise the covariance structure in repeated measures models, multivariate normal models and general multivariate mixed models may be in terms of the variance and correlation structures determined by separate sets of parameters. The robustness of coefficient ratios in regression models has important implications for model specification and interpretation, and has been studied by a number of authors. See, for instance, Brillinger (1983), Solomon (1984), Struthers and Kalbfleisch (1986), Skinner (1987), Li and Duan (1989) and Taylor (1989).

The estimation of underlying distributions by empirical cumulant generating functions is studied in depth by Hall and Yao (2002) and the estimation by cumulants by Cox and Hall (2002).

6.11 Further results and exercises

1. By a formula of Sheppard (1898), $\Phi_2(0,0;\rho) = 1/4 + \sin^{-1}(\rho)/(2\pi)$. Hence show that on the basis of only exceedances over the mean and equations (6.3) to (6.5) the parameter ρ can be estimated; the technique is called tetrachoric correlation. A test of adequacy of the originating model could be based on a comparison of the tetrachoric correlation

and the ordinary product-moment correlation of the continuous measurements. How could significance be assessed?

Appendix

A.1 Fitting separate logistic regressions to the ANZICS data

We outline here our application of the adjusted maximum likelihood R function `brlr` for biased reduced logistic regression used in Example 4.1: *Australian intensive care outcomes*. The function was written by D. Firth and is available from CRAN at `http://cran.r-project.org`.

Separate logistic regressions were fitted to each of the 109 hospitals in the ANZICS database, for which separately estimated regression coefficients and covariance matrices were obtained.

The response variable, here `diedhos`, must be a two-level factor (usually 1=dead, 0=alive). The individual covariance matrices are output to a three-dimensional array, and the regression coefficients to a 109×6 matrix: the matrix XX retains the structure of the individual covariance matrices.

```
> library(MASS)
> source("brlr.R")
> options(warn=1)
> beta <- array(dim=c(109,6))
> XX <- array(dim=c(6,6,109))
> for (i in 1:length(unique(siteid))){
+   print(i)
+   my.model <- brlr(diedhos~ apache +
+   Ddays*diffdays + apache:Ddays, br=TRUE,
+   subset=siteid==unique(siteid)[i])
+   beta[i,] <- my.model$coefficients
+   XX[,,i] <- vcov(my.model)
}
```

In addition, the convergence of each logistic regression fit was checked by including `my.model$convergence` in the loop and confirming it took the logical value TRUE in each case.

A description of the problem of separation effects in logistic regression and the R `brlr` function provided to overcome them are given in the Bibliographic notes and Computational/software notes for Chapter 4.

References

Airy, G.B. (1861). *On the algebraic and numerical theory of errors of observation and the combination of observations.* London: McMillan.

Albert, A. & Anderson, J.A. (1984). On the existence of maximum likelihood estimates in logistic regression. *Biometrika* **71**, 1–10.

Alizadeh, A.A, Eisen, M.B., Davis, R.E., Ma, C., Lossos, I.S., Rosenwald, J.C., Boldrick, J.C., Sabet, H., Tran, T., Yu, X., Powell, J.I., Yang, L., Marti, G.E., Moore, T., Hudson, J. Jr., Lu, L., Lewis, D.B., Tibshirani, R., Sherlock, G., Chan, W.C., Greiner, T.C., Weisenburger, D.D., Armitage, J.O., Warnke, R., Levy, R., Wilson, W., Grever, M.R., Byrd, J.C., Botstein, D., Brown, P.O. & Staudt, L.M. (2000). Distinct types of diffuse large B-cell lymphoma identified by gene expression profiling. *Nature* **403**, 503–511.

Altman, D.G., Lausen, B., Sauerbrei, W. & Schumacher, M. (1994). Dangers of using 'optimal' cutpoints in the evaluation of prognostic factors. *Journal of National Cancer Institute* **86**, 829-35.

Anderson, R.L. (1975). Designs and estimators for variance components. In *A Survey of Statistical Design and Linear Models*, J.N. Srivastava (Ed.). Amsterdam: North-Holland.

Anderson, R.L. & Bancroft, T.A. (1952). *Statistical theory in research.* New York: McGraw-Hill.

Andrews, D.F. & Stafford, J.E.S. (1999). *Symbolic computation for statistical inference.* Oxford: Oxford University Press.

Anscombe, F.J. (1950). Sampling theory of the negative binomial and log series distributions. *Biometrika* **37**, 358–382.

Azzalini, A. (1996). *Statistical inference based on the likelihood.* London: Chapman & Hall.

Bainbridge, T.R. (1965). Staggered nested designs for estimating variance components. *Industrial Quality Control* **22**, 12–20.

Balding, D.J., Bishop, M. & Cannings, C. (Eds.) (2001). *Handbook of Statistical Genetics.* Chichester, U.K.: Wiley.

Bammer, G. & McDonald, D.N. (1994). Report on a workshop on trial evaluation. In *Issues for designing and evaluating a 'heroin trial'*, NCEPH Working Paper 8. Canberra: The Australian National University.

Barndorff-Nielsen, O.E. & Cox, D.R. (1988). *Asymptotic techniques for use in statistics.* London: Chapman & Hall.

Barndorff-Nielsen, O.E. & Cox, D.R. (1994). *Inference and asymptotics.* London: Chapman & Hall.

Bartlett, M.S. (1937). Properties of sufficiency and statistical tests. *Proceedings of Royal Society (London) A* **160**, 268–282.

Birnbaum, A. (1968). Chapters in Lord, F.M. & Novick, M.R. (Eds.), *Statistical theory of mental test scores*. Reading, MA: Addison-Wesley.

Bolstad, B., Dudoit, S. & Yang, Y.H. (November 2001). sma v. 0.5.6, http://www.stat.berkeley.edu/users/terry/zarray/Software/smacode.html.

Booth, J.G. & Hobert, J.P. (1998). Standard errors of prediction in generalized linear mixed models. *Journal of American Statistical Association* **93**, 262–272.

Boscardin, W.J., Taylor, J.M.G. & Law, N. (1998). Longitudinal models for AIDS marker data. *Statistical Methods in Medical Research* **7**, 13–27.

Breslow, N.E. (2000). Whither PQL? *Second Seattle Symposium in Biostatistics.* Notes of conference lecture, November 18–21.

Breslow, N.E. & Clayton, D.G. (1993). Approximate inference in generalized linear mixed models. *Journal of American Statistical Association* **88**, 9–25.

Breslow, N.E. & Lin, X. (1995). Bias correction in generalized linear mixed models with a single component of dispersion. *Biometrika* **82**, 81–91.

Brillinger, D.R. (1983). A generalized linear model with 'Gaussian' regressor variables. In *A Festschrift for Erich Lehmann*, P.J. Bickel, K.A. Doksum and J.L. Hodges (Eds.), pp. 97–114. Belmont, CA: Wadsworth.

Brookmeyer, R. & Gail, M.H. (1988). A method of obtaining short-term projections and lower bounds on the size of the AIDS epidemic. *Journal of American Statistical Association* **83**, 301–308.

Brown, P.O. & Botstein, D. (1999). Exploring the new world of the genome with DNA microarrays. In *The Chipping Forecast* **21**, 33–37. Supplement to *Nature Genetics*, January 1999.

Bulmer, M.G. (1980). *The mathematical theory of quantitative genetics*. Oxford: Oxford University Press.

Carpenter, L.M., Maconochie, N.E.S., Roman, E. & Cox, D.R. (1997). Examining associations between occupation and health by using routinely collected data. *Journal of Royal Statistical Society A* **160**, 507–521.

Carroll, R.J., Ruppert, D. & Stefanski, L.A. (1995). *Measurement error in nonlinear models*. London: Chapman & Hall.

Clayton, D.G. (1996). Generalized linear mixed models. In *Markov chain Monte Carlo in practice*, Gilks, W.R., Richardson, S. and Spiegelhalter, D.J. (Eds.). London: Chapman & Hall.

Cochran, W.G. (1954). The combination of estimates from different experiments. *Biometrics* **10**, 101–129.

Commenges, D. & Jacqmin-Gadda, H. (1997). Generalized score test of homogeneity based on correlated random effects models. *Journal of Royal Statistical Society B* **59**, 157–171.

Cornfield, J. & Tukey, J.W. (1956). Average values of mean squares in factorials. *Annals of Mathematical Statistics* **27**, 907–949.

Cox, D.R. (1966). Some procedures connected with the logistic qualitative response curve. *Research papers in statistics; essays in honour of J. Neyman's 70th birthday*. F.N. David (Ed.), 55–71. Chichester: Wiley.

Cox, D.R. (1982). Combination of data. In *Encyclopedia of statistical sciences* S. Kotz and N.L. Johnson (Eds.), **2**, 45–53. New York: Wiley.

Cox, D.R. (1984). Effective degrees of freedom and the likelihood ratio test. *Biometrika* **71**, 487–493.

Cox, D.R. (1998). Components of variance: a miscellany. *Statistical Methods in Medical Research* **7**, 3–12.

Cox, D.R. & Hall, P. (2002). Estimation in a simple random effects model with nonnormal distributions. *Biometrika* **89**, to appear.

Cox, D.R. & Reid, N. (1987). Parameter orthogonality and approximate conditional inference (with discussion). *Journal of Royal Statistical Society B* **49**. 1–39.

Cox, D.R. & Solomon, P.J. (1986). Analysis of variability with large numbers of small samples. *Biometrika* **73**, 543–554.

Cox, D.R. & Solomon, P.J. (1988). On testing for serial correlation in large numbers of small samples. *Biometrika* **75**, 145–148.

Cox, D.R. & Wermuth, N. (1996). *Multivariate dependencies: models. analysis and interpretation.* London: Chapman & Hall.

Daniels, H.E. (1938). Some problems of interest in wool research. *Supplement of Journal of Royal Statistical Society* **5**, 89–128.

Daniels, H.E. (1939). The estimation of components of variance. *Supplement of Journal of Royal Statistical Society* **6**, 186–197.

De Angelis, D., Gilks, W.R. & Day. N.E. (1998). Bayesian projection of the acquired immune deficiency syndrome epidemic. *Journal of Royal Statistical Society C* **47**, 449–498.

Diggle, P.J., Liang, K.-Y., & Zeger, S.L. (1994). *Analysis of longitudinal data.* Oxford: Oxford University Press.

Duda, R.O. & Hart, P.E. (1973). *Pattern classification and scene analysis.* New York: Wiley.

Dudoit, S., Yang, Y.H., Speed, T.P. & Callow, M.J. (2002). Statistical methods for identifying differentially expressed genes in replicated cDNA microarray experiments. *Statistica Sinica,* **12**, 111–140.

Efron, B., Tibshirani, R., Storey, J.D. & Tusher. V. (2001). Empirical Bayes analysis of a microarray experiment. *Journal of American Statistical Association* **96**, 1151–1160.

Ehrenberg, A.S.C. (1950). The unbiased estimation of heterogenous error variances. *Biometrika* **7**, 347–357.

Eisen, M.B. & Brown, P.O. (1999). DNA arrays for analysis of gene expression. *Methods in Enzymology* **303**, 179–205.

Eisenhart, C. (1947). The assumptions underlying the analysis of variance. *Biometrics* **47**, 1–21.

Elston, R.C., Olson, J. & Palmer, L. (Eds.) (2002). *Biostatistical genetics and genetic epidemiology.* Chichester: Wiley.

Engel, B. & Keen, A. (1994). A simple approach for the analysis of generalized linear mixed models. *Statistica Neerlandica* **48**, 1–22.

Fahrmeir, L. & Tutz, G. (1994). *Multivariate statistical modelling based on generalized linear models.* New York: Springer.

Firth, D. (1993). Bias reduction of maximum likelihood estimates. *Biometrika* **80**, 27–38. Amendment: *Biometrika* (1995) **82**, 667–668.

Fisher, R.A. (1950). The significance of departures from expectations in a Poisson series. *Biometrics* **6**, 17–24.

Gilks, W., Richardson, S., & Spiegelhalter, D.J. (1996). *Markov chain Monte Carlo in practice*. London: Chapman & Hall.

Glonek, G.V.F. & Solomon, P.J. (2002). Factorial designs for microarray experiments. Preprint; http://www.maths.adelaide.edu.au/MAG.

Goldsmith, C.H. & Gaylor, D.W. (1970). Three stage nested designs for estimating variance components. *Technometrics* **12**, 487–498.

Goldstein, H. (1995). *Multilevel statistical models*. New York: Wiley.

Green, P.J. (1987). Penalized likelihood for general semi-parametric regression models. *International Statistical Review* **55**, 245–259.

Green, P.J. (2001). A primer on Markov chain Monte Carlo. In *Complex stochastic systems*, O.E. Barndorff-Nielsen, D.R. Cox and C. Klüppelberg (Eds.), pp. 1–62. Boca Raton: Chapman & Hall.

Green, J., Banks, E., Berrington, A., Darby, S., Deo, H. & Newton, R. (2000). N-acetyltransferase 2 and bladder cancer: an overview and consideration of the evidence for gene-environment interaction. *British Journal of Cancer* **83**, 412–417.

Greenwood, M. & Yule, G.U. (1920). An enquiry into the nature of frequency distributions representative of multiple happenings with particular reference to the occurrence of multiple attacks of disease or of repeated accidents. *Journal of Royal Statistical Society* **83**, 255–279.

Guo, S.-W. (1998). *Polygenic inheritance. Encyclopedia of Biostatistics*, P. Armitage and T. Colton (Eds.), pp. 3418–3424.

Hall, P. & Yao, Q. (2002). Inference in components of variance models with low replication. Submitted.

Hartley, H.O. & Rao, J.N.K. (1967). Maximum-likelihood estimation for the mixed analysis of variance model. *Biometrika* **54**, 93–108.

Hastie, T., Tibshirani, R., Eisen, M. B., Alizadeh, A., Levy, R., Staudt, L., Chan, W. C., Botstein, D. & Brown, P. (2000). 'Geneshaving' as a method for identifying distinct sets of genes with similar expression patterns. *Genome Biology* **1**(2): research0003.1-0003.21.

Hastings, W.K. (1970). Monte Carlo sampling methods using Markov chains and their applications. *Biometrika* **57**, 97–109.

Hedges, L.V. & Olkin, I. (1985). *Statistical methods for meta-analysis*. Orlando, FL: Academic Press.

Henderson, C.R. (1953). Estimation of variance and covariance components. *Biometrics* **9**, 226–252.

Henderson, C.R. (1975). Best linear unbiased estimation and prediction under a selection model. *Biometrics* **31**, 423–447.

Hopper, J.L. (1993). Variance components for statistical genetics: applications in medical research to characteristics related to diseases and health. *Statistical Methods in Medical Research* **2**, 199–223.

Hunt, C.H. & Solomon, P.J. (1999). On investigating mean–variance relationships in general component of variance models. Preprint.

Hurzurbazar, V.S. (1950). Probability distributions and orthogonal parameters. *Proceedings of Cambridge Philosophical Society* **46**, 241–284.

Ihaka, R. & Gentleman, R. (1996) *R*: A language for data analysis and graphics. *Journal of Computational and Graphical Statistics* **5**, 299–314.

Irwin, J.O. & Kendall, M.G. (1943–45). Sampling moments of moments for a finite population. *Annals of Eugenics* **12**, 138–142.

IPPPSH Collaborative Group (1984). The International Prospective Primary Prevention Study in Hypertension (IPPPSH): objectives and methods. *European Journal of Clinical Pharmacology* **27**, 379–391.

IPPPSH Collaborative Group (1985). Cardiovascular risk and risk factors in a randomized trial of treatment based on the beta-blocker oxprenolol: the International Prospective Primary Prevention Study in Hypertension. *Journal of Hypertension* **3**, 379–392.

Isham, V.S. (1988). Estimation of the incidence of HIV infection: the back projection method. Appendix in *Short-term projection of HIV infection and AIDS*. London: HMSO.

Isham, V.S. (1989). Estimation of the incidence of HIV infection. *Philosophical Transactions of Royal Society (London) B* **325**, 113–121.

Jin, W., Riley, R.M., Wolfinger, R.D., White, K.P., Passador-Gurgel, G. & Gibson, G. (2001). The contributions of sex, genotype and age to transcriptional variance in *Drosophila melanogaster*. *Nature Genetics* **29**, 389–395.

Kalbfleisch, J.D. & Sprott, D.A. (1970). Application of likelihood methods to models involving large numbers of parameters (with discussion). *Journal of Royal Statistical Society B* **32**, 175–208.

Ke, C. & Wang, Y. (2001). Semiparametric nonlinear mixed-effects models and their application (with discussion). *Journal of American Statistical Association* **96**, 1272–1298.

Kerr, M.K., Martin, M. & Churchill, G.A. (2000). Analysis of variance for gene expression microarray data. *Journal of Computational Biology* **7**, 819–837.

Kerr, M.K. & Churchill, G.A. (2001). Experimental design for gene expression microarrays. *Biostatistics* **2**, 183–201.

Khuri, A.I. (2000). Designs for variance component estimation: past and present. *International Statistical Review* **68**, 311–322.

Khuri, A.I. & Sahai, H. (1985). Variance components analysis: a selective bibliography and survey. *International Statistical Review* **53**, 279–300.

Knaus, W., Draper, E.A., Wagner, D.P. & Zimmerman, J.E. (1985). APACHE II: a severity of disease classification system. *Critical Care Medicine* **13**, 818–829.

Laird, N. (1978). Empirical Bayes methods for two-way contingency tables. *Journal of American Statistical Association* **65**, 581–590.

Lancaster, T. & Nickell, S. (1980). The analysis of re-employment probabilities for the unemployed (with discussion). *Journal of Royal Statistical Society A* **143**, 141-135.

Law, G.R., Cox, D.R., Maconochie, N.E.S., Simpson, J., Roman, E. & Carpenter, L.M. (2001). Large tables. *Biostatistics* **2**, 163–171.

Lee, Y. & Nelder, J.A. (1996). Hierarchical generalized linear models (with discussion). *Journal of Royal Statistical Society B* **58**, 619–678.

Lee, Y. & Nelder, J.A. (2001). Hierarchical generalized linear models: a synthesis of generalized linear models, random-effect models and structural dispersions. *Biometrika* **88**, 987–1006.

Li, K.C. & Duan, N. (1989). Regression analysis under link violation. *Annals of Statistics* **17**, 1009–1052.

Lin, X. & Breslow, N.E. (1996). Analysis of correlated binomial data in logistic-normal models. *Journal of Statistical Simulation and Computation* **55**, 130–146.

Lönnstedt, I., Grant, S., Begley, G. & Speed, T.P. (2001) Microarray analysis of two interacting treatments: a linear model and trends in expression over time. Technical Report, Department of Mathematics, Uppsala University, Sweden.

Lönnstedt, I. & Speed, T.P. (2002). Replicated microarray data. *Statistica Sinica* **12**, 31–46.

McCullagh, P. (2000). Invariance and factorial models (with discussion). *Journal of Royal Statistical Society B* **62**, 209–256.

McCullagh, P. & Nelder, J.A. (1989). *Generalized linear models.* 2nd edition. London: Chapman & Hall.

McCulloch, C.E. & Searle, S.R. (2001). *Generalized, linear, and mixed models.* New York: Wiley.

Metropolis, N., Rosenbluth, A.W., Rosenbluth, M.N., Teller, A.H. & Teller, E. (1953). Equations of state calculations by fast computing machines. *Journal of Chemical Physics* **21**, 1087–1091.

Myles, J. & Clayton, D.G. (2001). *An R package for estimating Bayesian generalized linear mixed models by Gibbs sampling.* Available from `http://cran.r-project.org`.

Nelder, J.A. (1977). A reformulation of linear models (with discussion). *Journal of Royal Statistical Society A* **140**, 48–77.

Olsen, M.K. & Shafer, J.L. (2001). A two-part random-effects model for semi-continuous longitudinal data. *Journal of American Statistical Association* **96**, 730–745,

Patterson, H.D. & Thompson, R. (1971). Recovery of inter-block information when block sizes are unequal. *Biometrika* **58**, 545–554.

Pinheiro, J.C. & Bates, D.M. (1995). Approximations to the log-likelihood function in the nonlinear mixed-effects model. *Journal of Computational and Statistical Graphics* **4**, 12–35.

Pinheiro, J.C. & Bates, D.M. (2000). *Mixed-effects models in S and S-PLUS.* Springer: New York.

Rabe-Hesketh, S., Pickles, A. & Skondral, A. (2001). *GLLAMM Manual.* Technical Report 2001/01. Department of Biostatistics and Computing, Institute of Psychiatry, King's College, University of London. `http://www.iop.kcl.ac.uk/iop/departments/biocomp/programs/gllamm.html`.

Rabe-Hesketh, S., Skondral, A. & Pickles, A. (2002). Reliable estimation of generalized linear mixed models using adaptive quadrature. *The Stata Journal* **2**, 1–21.

Rao, C.R. & Kleffe, J. (1988). *Estimation of variance components and applications.* Amsterdam: North-Holland.

Rao, P.S.R.S. (1997). *Variance component estimation.* London: Chapman & Hall.

Reeves, G.K., Cox, D.R., Darby, S.C. & Whitley, E. (1998). Some aspects of measurement error in explanatory variables for continuous and binary regression models. *Statistics in Medicine* **17**, 2157–2177.

Ripley, B.D. (1996). *Pattern recognition and neural networks.* Cambridge: Cambridge University Press.

Robinson, G.K. (1991). That BLUP is a good thing: the estimation of random effects (with discussion). *Statistical Science* **6**, 15–51.

Rocke, D.M. & Lorenzato, S. (1995). A two-component model for measurement error in analytical chemistry. *Technometrics* **37**, 176–184.

Rocke, D.M. & Durbin, B. (2001). A model for measurement error for gene expression arrays. *Journal of Computational Biology* **8**, 557–569.

Scheffé, H. (1959). *The analysis of variance.* New York: Wiley.

Schena, M. (Ed.) (1999). *DNA microarrays: a practical approach.* New York: Oxford University Press.

Schena, M. (Ed.) (2000). *Microarray biochip technology.* Natick. MA: Eaton Publishing.

Searle, S.R., Casella, G. & McCulloch. C.E. (1992). *Variance components.* New York: Wiley.

Sheppard, W.F. (1898). On the geometric treatment of the 'normal curve' of statistics with particular reference to correlation and the theory of errors. *Proceedings Royal Society (London)* **62**. 170–173.

Shun, Z. (1997). Another look at the salamander mating data: a modified Laplace approximation approach. *Journal of American Statistical Association.* **92**. 341–349.

Skellam, J.G. (1948). A probability distribution derived from the binomial by regarding the probability of success as variable between sets of trials. *Journal of Royal Statistical Society B* **10**. 257–261.

Skinner, C.J. (1987). Discussion on *Parameter orthogonality and approximate conditional inference* (by D.R. Cox and N. Reid). *Journal of the Royal Statistical Society B* **49**, 24.

Skoog, D.A., West, D.M. & Holler. J. (1995). *Fundamentals of analytical chemistry* 7th ed. Philadelphia: Saunders Golden.

Snedecor, G.W. & Cochran, W.G. (1967). *Statistical methods.* 6th ed. Ames. IA: Iowa State College Press.

Snijders, T. & Bosker, R. (1999). *Multilevel Analysis: an introduction to basic and advanced multilevel modelling.* London: Sage.

Solomon, P.J. (1984). Effect of misspecification of regression models in the analysis of survival data. *Biometrika* **71**. 291–198. Amendment (1986). **73**. 245.

Solomon, P.J. (1985). Transformations for components of variance and covariance. *Biometrika* **72**, 233–239.

Solomon, P.J. (1989). On components of variance and modelling exceedances over a threshold. *Australian Journal of Statistics* **31**, 18–24.

Solomon, P.J. (Ed.) (1998). Five papers on variance components in medical research. *Statistical Methods in Medical Research* **7**, 1–84.

Solomon, P.J. & Cox, D.R. (1992). Nonlinear component of variance models. *Biometrika* **79**, 1–11.

Solomon, P.J., Thompson, E.A. & Rissanen. A. (1983). The inheritance of height in a Finnish population. *Annals of Human Biology* **10**, 247–256.

Solomon, P.J. & Taylor, J.M.G. (1999). Orthogonality and transformations in variance components models. *Biometrika* **86**, 289–300.

Solomon, P.J. & Wilson, S.R. (2001). Statistical modelling and prediction associated with the HIV/AIDS epidemic. *The Mathematical Scientist* **26**. 87–102.

Spiegelhalter, D., Thomas, A., Best, N. & Gilks, W. (1996). *BUGS 0.5 Bayesian analysis using Gibbs sampling. Manual (*version ii*).* Cambridge: MRC Biostatistics Unit. http://www.mrc-bsu.cam.ac.uk/bugs/documentation/contents.html.

Spiegelhalter, D., Abrams, K. & Myles, J. (2002). *Bayesian approaches to clinical trials and health-care evaluation.* Chichester, UK: Wiley.

StataCorp. (2001). *Stata Statistical Software: Release 7.* College Station, TX.

Stiratelli, R., Laird, N. & Ware, J. (1984). Random effects models for serial observations with binary responses. *Biometrics* **40**, 961–971.

Struthers, C.A. & Kalbfleisch, J.D. (1986). Misspecified proportional hazards models. *Biometrika* **73**, 363–369.

Sutton, A.J., Abrams, K.R., Jones, D.R., Sheldon, T.A. & Song, F. (2000). *Methods for meta-analysis in medical research.* Chichester: Wiley.

Sy, J.P., Taylor, J.M.G. & Cumberland,W.G. (1997). A stochastic model for the analysis of bivariate longitudinal AIDS marker data. *Biometrics* **53**, 542–555.

Taylor, J.M.G. (1989). On the cost of estimating the ratio of regression parameters after fitting a power transformation. *Journal of Statistical Planning and Inference* **21**, 223–230.

Taylor, J.M.G., Cumberland, W.G. & Meng, X. (1996). Components of variance models with transformations. *Australian Journal of Statistics* **38**, 183–191.

The Chipping Forecast (1999). *Nature Genetics* Supplement **21**, 1–60.

Tippett, L.H.C. (1931). *Methods of statistics.* London: Matthew Norgate.

Tukey, J.W. (1950). Some sampling simplified. *Journal of American Statistical Association* **45**, 501–519. Reprinted in *The collected works of John W. Tukey,* vol. 7. Pacific Grove: Wadsworth.

Venables, W.N. & Ripley, B.D. (1999). *Modern applied statistics with S-PLUS,* 3rd ed. New York: Springer.

Verbeke, G. & Molenberghs, G. (2000). *Linear mixed models for longitudinal data.* New York: Springer.

Wilk, M.B. & Gnanadesikan, R. (1964). Graphical methods for internal comparisons in multiresponse experiments. *Annals of Mathematical Statistics* **35**, 613–631.

Wolfinger, R.D., Gibson, G., Wolfinger, E., Bennett, L., Hamadeh, H., Bushel, P., Afshari, C. & Paules, R.S. (2001). Assessing gene significance from cDNA microarray expression data via mixed models. *Journal of Computational Biology* **8**, 625–637.

Yang, M., Goldstein, H. & Heath, A. (2000). Multilevel models for repeated binary outcomes: attitudes and voting over the electoral cycle. *Journal of Royal Statistical Society A* **163**, 49–62.

Yates, F. & Cochran, W.G. (1938). The analysis of groups of experiments. *Journal of Agricultural Science* **28**, 556–580.

Zeger, S.L. & Karim, M.R. (1991). Generalized linear models with random effects: a Gibbs sampling approach. *Journal of American Statistical Association* **86**, 79–86.

Zorn, M.E., Gibbons, R.D. & Sonzogni, W.C. (1997). Weighted least-squares approach to calculating limits of detection and quantification by modelling variability as a function of concentration. *Analytical Chemistry* **69**, 3069–3075.

REFERENCES

Zorn, M.E., Gibbons, R.D. & Sonzogni, W.C. (1999). Evaluation of approximate methods for calculating the limit of detection and the limit of quantification. *Environmental Science and Technology* **33**, 2291–2295.

Author index

Subject index